I0053650

# Ecosystemic Evolution Feeded by Smart Systems

## Special Issue Editor
Dino Giuli

MDPI • Basel • Beijing • Wuhan • Barcelona • Belgrade

**MDPI**

*Special Issue Editor*
Dino Giuli
University of Florence
Italy

*Editorial Office*
MDPI AG
St. Alban-Anlage 66
Basel, Switzerland

This edition is a reprint of the Special Issue published online in the open access journal *Future Internet* (ISSN 1999-5903) from 2016–2018 (available at: http://www.mdpi.com/journal/futureinternet/special_issues/Ecosystemic-Evolution-Smart-Systems).

For citation purposes, cite each article independently as indicated on the article page online and as indicated below:

Lastname, F.M.; Lastname, F.M. Article title. *Journal Name* **Year**, Article number, page range.

**First Edition 2018**

**ISBN 978-3-03842-829-9 (Pbk)**
**ISBN 978-3-03842-830-5 (PDF)**

# Table of Contents

# About the Special Issue Editor

**Dino Giuli** was a Full Professor of Telecommunications at the Department of Information Engineering (DINFO) in the University of Florence (IT) from 1986 until his retirement in 2016. He then became a scientific consultant for the same University. His research activities have been progressively devoted and empowered within Radar, Environmental Monitoring and Smart Systems, by promoting and exploiting an extended interdisciplinary and transdisciplinary approach. He is senior member of IEEE and was the Director of the Department of Electronics and Telecommunications at the University of Florence from 1999 until 2004. From 1987 to 1993 he was the President of the Centre for Information and Telematics Services of the University of Florence. From 1987 to 1993 he was a member of the technical committee of CINECA, the University National Council for automatic computing. Since 1996 until 2016 he was the Director of the interdisciplinary Ph.D. School in "Telematics and Information Society" at the University of Florence. He has been a scientific co-ordinator of, or participated in many national and international research projects pertaining to his fields of research activity.

*future internet*

MDPI

*Editorial*

# Ecosystemic Evolution Fed by Smart Systems    f

## Dino Giuli

Department of Information Engineering (DINFO), University of Florence, Via Santa Marta, 3, Florence 50139, Italy; dino.giuli@unifi.it

Received: 5 March 2018; Accepted: 6 March 2018; Published: 10 March 2018

Information Society is advancing along a route of ecosystemic evolution. ICT and Internet advancements, together with the progression of the systemic approach for enhancement and application of Smart Systems, are grounding such an evolution. The needed approach is therefore expected to evolve by increasingly fitting into the basic requirements of a significant general enhancement of human and social well-being, within all spheres of life (public, private, professional). This implies enhancing and exploiting the net-living virtual space, to make it a virtuous beneficial integration of the real-life space. Meanwhile, contextual evolution of smart cities is aiming at strongly empowering that ecosystemic approach by enhancing and diffusing net-living benefits over our own lived territory, while also incisively targeting a new stable socio-economic local development, according to social, ecological, and economic sustainability requirements. This territorial focus matches with a new glocal vision, which enables a more effective diffusion of benefits in terms of well-being, thus moderating the current global vision primarily fed by a global-scale market development view.

Basic technological advancements have thus to be pursued at the system-level. They include system architecting for virtualization of functions, data integration and sharing, flexible basic service composition, and end-service personalization viability, for the operation and interoperation of smart systems, supporting effective net-living advancements in all application fields. Increasing and basically mandatory importance must also be increasingly reserved for human–technical and social–technical factors, as well as to the associated need of empowering the cross-disciplinary approach for related research and innovation. The prospected eco-systemic impact also implies a social pro-active participation, as well as coping with possible negative effects of net-living in terms of social exclusion and isolation, which require incisive actions for a conformal socio-cultural development. In this concern, speed, continuity, and expected long-term duration of innovation processes, pushed by basic technological advancements, make ecosystemic requirements stricter. This evolution requires also a new approach, targeting development of the needed basic and vocational education for net-living, which is to be considered as an engine for the development of the related 'new living know-how', as well as of the conformal 'new making know-how'.

The papers of the special issue are significant contributions samples within the general ecosystemic view above highlighted. The first group of papers ([1–5]) pertain the focus on human and social factors through interdisciplinary and transdisciplinary approaches, which look for enhancing quality of life supported by the smart system environment.

The second group of papers ([6–9]) pertains to some relevant technological infrastructure enhancements, based on exploitation of the Internet of Things the 5G network paradigms.

Several contributed papers widely point out also the relevance of smart city context for grounding and developing the ecosystemic approach, through advancement of both intangible and tangible infrastructures. Contributions made by each paper is below outlined.

The first paper—by Martelli [1]—exposes and grounds an initiative point of view about new education needed for smart citizenship. Education has to cope indeed with citizen awareness lack. The paper's primary concern is about privacy and freedom requirements connected with the knowledge bases generated by the IT services. For such a purpose, an original education methodology is prospected in the paper, while also reporting and discussing related experiences made through some specifics use cases.

The second paper—by Angelini, Carrino, Khaled, Riva-Mossman, and Mugellini [2]—describes and discusses a new transdisciplinary research platform jointly developed by four universities to support co-creation of ICT based innovative products, services, and practices, to fit into user requirements of older adults. Such a platform is devised as the Senior Living Lab (SLL), by exploiting cooperative transdisciplinary support of designers economists, engineers, and healthcare professionals, as well as pro-active participation of end users (older adults). The paper describes the approach thus adopted for the lab's operation, as well reporting objectives and operational outcomes of some current projects concerned with healthy nutrition to cope with frailty, improved autonomous mobility, and social communication to prevent isolation.

The third paper—by Marti, Megens, and Hummels [3]—exploits an ecosystemic view which focuses on the needed advancements of "user-centered design" for social innovation. Paper focus is needed on advanced methodologies for transforming data-driven design towards data-enabled participatory and co-creative design. The new methodology purposely devised to design smart systems, called experimental design landscapes (EDL), is presented in this paper, together with two samples projects. Such projects are re4specively concerned with the following topics: human behavior change mediated by sensing technologies; social platform to sustain new processes of deliberative democracy.

The fourth paper—by Lettieri [4]—is a significant contribution on current advancements and perspectives of computational social sciences (CSS) for evolution of policy design and rule making within smart societies. While providing the needed review, the paper highlights and discusses promising scientific advancements, which can be prospected for CSS valuing. The primary focus is on the new mechanism needed for regulating both formulation and evolution towards a smart society.

The fifth paper—by Guidi, Miniati, Mazzolla, and Iadanza [5]—pertains to smart health systems application. This contribution is concerned with the use of a cloud platform to provide analytics-as-a-service (AaaS) tools, for heart disease continuous monitoring and fast detection. Application and experimentations are described in the paper, which have been developed through personalization of basic IBM Watson Analytics tools, while exploiting only electrocardiographic signal and heart rate variability, for monitoring and heart failure disease detection. Based on experimental outcomes, advantages and drawbacks of the cloud approach with respect to the usual static approach are discussed in the paper.

The sixth paper—by Carrino, Mugellini Khaled, Ouerhani, and Ehrensberger [6]—is a significant contribution concerned with advancements of Internet of Things for Urban Innovation (iNUIT), as a key support for smart city development. The paper's contributions are framed within a specific multi-year research program started in Switzerland. Reported research activities and results refers to two started projects which are included in such a program. The first project (smart crowd) is concerned with monitoring the crowd's movement to detect possible dangerous scenarios. Purposeful, real-time tracking is afforded through sensors which are available within smart-phones. The second project (OpEc) is concerned with exploitation of an Internet of Things approach, to implement dynamic street light management and control, aiming at street light energy saving. Shown experimental results of both projects point out efficacy of the adopted solutions.

The seventh paper—by Paradiso, Paganelli, and Giuli [7]—is contributing to the enhancements of "Internet of Things" approach aimed at energy consumption cost saving in residential places. The paper is specifically proposing and evaluating a new non-intrusive load monitoring (NILM) instrumental asset, keeping the needed capabilities for monitoring and control of energy consumption while reducing instrumental complexity and cost of the current load monitoring solutions. The proposed NILM solution is based on: (i) disaggregation of whole-house consumption into the single portions associate to each consuming device; (ii) exploiting information of users presence and hourly use of appliances; (iii) enabling a constructive behavior of final user for energy saving; (iv) addressing monitoring for total active power measurements, which can be sampled at a much lower frequency thus reducing their data set size. As shown in the paper, positive results have been obtained through simulations fed through and experimental open dataset.

The eighth paper—by Martini and Paganelli [8]—is concerned with emerging network technologies based on the paradigms of software-defined networks (SDN) and network function virtualization (NFV), which are aimed at enhancement of network operation flexibility and cost reduction. Such an evolution is particularly relevant for the current roadmap of the 5G network and for its impact in smart city contexts. In such a context, criticalities and limitations currently emerge for cross-virtualization and dynamic integration of network services among different network operators. The paper contribution is concerned with such a problem. A new approach is proposed and discussed, which is based joint exploitation of the service-oriented architecture (SOA) paradigm, to cope with the multi-operator virtual network integration, thus complementing the current NFV and SDN approaches. A new network architecture is thus worked out and discussed. Preliminary results of prototype implementation and testing activities are also presented, which highlights also benefits for network service providers.

The ninth paper—by Fantacci and Marabissi [9]—is concerned with advancements of telecommunications technologies for wireless communications. The topics faced pertain to the current roadmap of the 5G network, which are particularly important for related critical requirements arising within the smart city context. A specific contribution is indeed made about a new methodology to be adopted to improve usage of all potential spectrum resources. Resorting to cognitive radio technology is prospected and discussed, in order to support context-aware dynamic optimization of the spectrum usage and sharing. A review is made on such a subject, by exploiting two relevant new network paradigms: heterogeneous networks and Machine-to-Machine communications.

**Acknowledgments:** The guest editors wish to thank all the contributing authors, the professional reviewers for their precious help with the review assignments, and the excellent editorial support from the *Future Internet* journal at every stage of the publication process of this special issue.

**Conflicts of Interest:** The authors declare no conflict of interest.

## References

1. Martelli, C. A point of view on New Education for Smart Citizenship. *Future Internet* **2017**, *9*, 4. [CrossRef]
2. Angelini, L.; Carrino, S.; Khaled, O.A.; Riva-Mossman, S.; Mugellini, E. Senior Living Lab: An Ecological Approach to Foster Social Innovation in Ageing Innovation. *Future Internet* **2016**, *8*, 50. [CrossRef]
3. Marti, P.; Megens, C.; Hummels, C. Data Enabled Design for Social Change: Two Case Studies. *Future Internet* **2016**, *8*, 46. [CrossRef]
4. Lettieri, N. Computational Social Science, the Evolution of Policy Design and Rule Making in Smart Societies. *Future Internet* **2016**, *8*, 19. [CrossRef]
5. Guidi, G.; Miniati, R.; Mazzola, M.; Iadanza, E. Case Study: IBM Watson Analytics Cloud Platform as Analytic-as-aServbice System for Hearth Failure Early Detection. *Future Internet* **2016**, *8*, 32. [CrossRef]
6. Carrino, F.; Mugellini, E.; Abou Khaled, O.; Ouerhani, N.; Ehrensberger, J. iNuit: Internet of Things for Urban Innovation. *Future Internet* **2016**, *8*, 18. [CrossRef]
7. Paradiso, F.; Paganelli, F.; Giuli, D.; Capobianco, S. Context-Base Enertgy Disaggregation in Smart Homes. *Future Internet* **2016**, *8*, 4. [CrossRef]
8. Martini, B.; Paganelli, F. A Service-Oriented Approach for Dynamic Chaining of Virtual Network Functions over Multi-Provider Software-Defined Networks. *Future Internet* **2016**, *8*, 24. [CrossRef]
9. Fantacci, R.; Marabissi, D. Cognitive Spectrum Sharing: An Enabling Wireless Communication Technology for a Wide Use of Smart Systems. *Future Internet* **2016**, *8*, 23. [CrossRef]

*future internet*

MDPI

*Article*

# Context-Based Energy Disaggregation in Smart Homes

f

**Francesca Paradiso [1,*], Federica Paganelli [2], Dino Giuli [1] and Samuele Capobianco [2]**

[1]   Department of Information Engineering, University of Firenze, via S. Marta 3, 50139 Firenze, Italy;
     dino.giuli@unifi.it
[2]   Consorzio Nazionale Interuniversitario per le Telecomunicazioni (CNIT) Research Unit at the University of
     Firenze, via S. Marta 3, 50139, Firenze, Italy;
     federica.paganelli@unifi.it (F.P.); samuele.capobianco@stud.unifi.it (S.C.)
*   Correspondence: francesca.paradiso@unifi.it; Tel.: +39-055-275-8597; Fax: +39-055-275-8570

Academic Editor: Jose Ignacio Moreno Novella
Received: 26 November 2015; Accepted: 14 January 2016; Published: 27 January 2016

**Abstract:** In this paper, we address the problem of energy conservation and optimization in residential environments by providing users with useful information to solicit a change in consumption behavior. Taking care to highly limit the costs of installation and management, our work proposes a Non-Intrusive Load Monitoring (NILM) approach, which consists of disaggregating the whole-house power consumption into the individual portions associated to each device. State of the art NILM algorithms need monitoring data sampled at high frequency, thus requiring high costs for data collection and management. In this paper, we propose an NILM approach that relaxes the requirements on monitoring data since it uses total active power measurements gathered at low frequency (about 1 Hz). The proposed approach is based on the use of Factorial Hidden Markov Models (FHMM) in conjunction with context information related to the user presence in the house and the hourly utilization of appliances. Through a set of tests, we investigated how the use of these additional context-awareness features could improve disaggregation results with respect to the basic FHMM algorithm. The tests have been performed by using Tracebase, an open dataset made of data gathered from real home environments.

**Keywords:** energy; smart grid; smart home; metering; energy efficiency; Gaussian mixture models; Factorial Hidden Markov Models; energy disaggregation; context awareness; non intrusive load monitoring

---

## 1. Introduction

Achieving greater energy efficiency through ICT has become an increasingly relevant research topic in the last decade. With the steady rise in consumption and the decreasing availability of energy resources, a remarkable slowing down in energy wasting, especially through the widespread adoption of energy saving solutions, is increasingly targeted.

It is expected that proper use of ICT (e.g., sensing, processing and actuation capabilities) would facilitate the achievement of this objective, in both domestic and industrial domains. The private home domain especially absorbs a non-negligible percentage of the energy demand. Indeed, domestic consumptions represent approximately one third of the whole energy usage in the European Union [1] as well as in the United States [2].

Several studies on domestic consumption habits [3,4], have shown that often users are not aware of how much energy is consumed by the devices they use. It has been recognized [5] that this may impair the understanding and adoption of energy saving behaviors. In other words, if the user were informed about how much a specific device affects total consumption, he might change his behavior in order to save energy as well as money.

Hence, in this context, the introduction of Load Monitoring techniques, which support the continuous monitoring of electricity consumption and the consequent analysis of measured data, can also help in providing end-users with information and suggestions for improving their consumption behavior.

Load monitoring techniques can be grouped into three categories:

1. Non-Intrusive Load Monitoring (NILM) [6]: NILM refers to a family of techniques whose purpose is to derive the power consumption of a specific device from the whole-house consumption profile.
2. Hardware-based sub-metering: this technique is based on the deployment of a distributed system of low-cost metering devices (*i.e.*, smart plugs attached onto household appliances) connected through a wireless and/or wired network infrastructure to a data collection module.
3. Adoption of smart appliances: this approach relies on the use of household appliances enhanced with sensing, processing and communication capabilities that can remotely be controlled and configured.

Although the adoption of smart appliances would facilitate the user in implementing cost and energy actions, this approach is not likely to be put in place in the short term. Moreover, only a subset of devices are usually available as "smart appliances", such as TVs, dishwashers, and ovens.

On the other side, smart plugs can be attached to almost any type of device. However, this approach can be resource demanding since a fine grained monitoring would require the use of a relevant number of smart plugs. In addition to the required financial commitment, the physical deployment might not be easy for fixed appliances (*i.e.*, washing machine, dishwasher, refrigerator, *etc.*) or the user may be bothered by the obligation to constantly attach a smart plug to every portable device (*i.e.*, hair dryer, phone charger, laptop, *etc.*).

On the other hand, NILM approaches which are based on whole-house consumption information can be easily deployed by leveraging existing and widely adopted smart meters. Several NILM algorithms have been proposed in literature [7] to disaggregate the output of smart meters. Most of them need monitoring data sampled at high frequency (at least 1 GHz frequency). In real-world scenarios, this assumption may be resource demanding whether the computation is performed locally in a Home Energy Management System (where data storage and processing resource-intensive tasks are performed) or in a remote server (since a high amount of data has to be transferred).

In this paper, we propose an NILM approach that relaxes the requirements on monitoring data since it uses total active power measurements gathered at low frequency (about 1 Hz). On one hand, this design choice has the advantage of allowing the use of low-cost metering devices. On the other hand, low-frequency measurements contain less information useful for load disaggregation than high frequency ones. To cope with this issue, in this paper, we enhance state of the art disaggregation approaches based on Factorial Hidden Markov Models (FHMM) [8] with the use of context information, *i.e.*, information that can be gathered by home sensors on relevant events in the domestic environment to improve the accuracy of the disaggregation algorithm.

Our context-based energy disaggregation approach uses probabilistic models representing the appliances consumption behavior. More specifically, we adopted the additive Factorial Hidden Markov Model (FHMM) [9], where the observed variables represent the aggregated power consumption profile, while the hidden variables represent the states of appliances. Context information (namely user consumption patterns and users presence in a room) is exploited to vary the state transition probabilities of device models in order to improve the accuracy of results.

Moreover, the proposed approach has been tested using data gathered from real home environments and made available as an open dataset by the Technische Universität Darmstadt (*i.e.*, Tracebase [10]). In our opinion, this choice may be scientifically relevant since it eases the comparison of results with future work and encourages further improvements.

The paper is structured as follows: in Sections 2 and 3, we discuss Background and Related Work, respectively. Section 4 describes the disaggregation algorithm focusing on our context conditioning approach. In Section 5, we describe the testing activities and discuss related results. Section 6 concludes the paper with final considerations.

## 2. Background

This section provides background information on load monitoring and appliance profiling.

Appliance Profiling refers to the observation of an electronic device's consumption behavior in order to extract all the features that could characterize it in detail. It consists of defining a set of relations between the working states of an appliance and the energy that it consumes [11]. Thanks to the knowledge of these characterizing features, a monitoring system would be able to analyze the output of a meter and recognize the appliance(s) in use.

As suggested by Hart [6] and Zeifman and Roth [12], depending on their power profile, home appliances can be divided into four main categories:

1. *Permanent consumer devices.* Devices that are permanently on and are characterized by an almost constant power trace (e.g., smoke alarms, telephones, *etc.*).
2. *On-off appliances.* Appliances that can be modeled with on/off states (e.g., lamp, toaster, *etc.*).
3. *Finite State Machines (FSM) or Multistate devices.* Devices that pass through several switching states. An operation cycle can thus be represented through a Finite State Machine and can be repeated on a daily or weekly basis. Examples are a washing machine, a dishwasher, a clothes dryer, *etc.*
4. *Continuously variable consumer devices.* Devices that are characterized by a variable non-periodic power trace. Examples of such appliances include notebook and vacuum cleaners.

Furthermore, in order to characterize the behavior of an appliance, a minimal set of three power mode states can be defined [13]:

- *Active*: the appliance is fully operational; the trend of the power consumption trace depends on the specific appliance.
- *Stand by*: the appliance is turned off, but some activities continue to run. The power consumption trace is zero, except for some sporadic low consumption samples.
- *Disconnected*: the device is disconnected from the electric network.

A further classification can be made by considering the type of device load: resistive, inductive or capacitive load. This differentiation is related to the typology of device internal circuits and strongly influences its power consumption profile. The Active Power is the real part of the Apparent Power complex equation; it represents the amount of energy consumed by an appliance during its ON period. Since the Apparent Power is the product between the current and voltage effective values, then a current/voltage shifting causes a variation in the power transferred to the appliance. This variation can be detected through the analysis of the Reactive Power, the imaginary part of the Apparent Power equation, which represents the amount of power absorbed by inductive/capacitive elements and therefore not exploited by the load. As stated in [13] "the larger the current/voltage shift the grater the imaginary component" and, consequently, the lower the active power is transferred to the appliance. Therefore, the types of component that can be found in a device can be distinguished as follows:

- Inductive type: affects the power consumption by shifting the alternate voltage with respect to the alternate current (e.g., washing machine).
- Capacitive type: affects the power consumption by shifting the alternate current with respect to the alternate voltage (e.g., rechargeable battery).
- Resistive type: shows no shift of current and voltage; if the appliance is a pure resistive type, the current and voltage waveforms will always be in phase and the imaginary part (reactive power) of the complex apparent power is zero (e.g., toaster).

An appliance profile, also mentioned as "appliance signature" or "appliance fingerprint", is thus composed by several characteristics which can help to identify that specific device (e.g., real power, maximum power value, waveform shape, ON period duration, *etc.*).

A refrigerator power trace, for example, presents a periodic pattern whose periods depend on the overcoming of an internal temperature threshold manually or automatically set. This appliance is always connected to the electric network. A washing machine is switched on to perform a washing program and presents a consumption cycle over a specific time interval. Instead, an LCD television, even if it causes occasional consumption peaks due to sequences of very clear pictures, presents an almost uniform power trace; a microwave oven has typically a minute-usage and presents uniform peaks of high consumption. A coffee maker consumes less than the microwave oven, but they have a similar behavior: long periods of inactivity interspersed with short duration periods of almost uniform consumption.

## 3. Related Work

Non-Intrusive Load Monitoring (NILM) [6] is a research field that has been studied for more than 20 years and has recently received particular attention for its expected benefits in energy monitoring and conservation policies. As mentioned above, NILM techniques aim at disaggregating consumption data, obtained from a metering device (e.g., smart meter) connected to the electric network, in order to identify the energy consumed by single devices in private households. The Non-Intrusive qualification refers to the fact that these approaches do not require the use of metering hardware dedicated to each single appliance; this implies a shorter installation time and negligible user involvement.

The first NILM method, developed by the Hart's working group [6], was based on the continuous monitoring of the active and reactive power measured at the electric meter. This method allowed detecting only the status change of bi-state (ON-OFF) devices and those modeled by Finite State Machine (FSM). The obtained results showed poor accuracy, mainly because of the poor reliability and precision of the measuring instruments that were available in the early 90s.

### 3.1. Features

The NILM state of the art presents numerous works that differ in the type of features employed. Technological progress has made possible the refinement of metering hardware and allowed managing bigger quantity of data collected at ever higher frequencies. Nowadays, there are a lot available metering solutions with a configurable sampling rate. With low-frequency rates, we refer to sampling rates up to 1 kHz, which allow gathering steady-state features as opposed to those known as high-frequency (up to 100 MHz), at which even the transient-state features can be detected [7].

#### 3.1.1. Low Sampling Rate

The choice to work with low sampling rates allows for analyzing steady-state features and provides several advantages from the economic point of view; the hardware required to collect these features has, in fact, a relatively low cost. One of the most investigated feature is the Real Power, which has been defined in the previous section. Several works [14–17] have tried to use this unique feature to perform disaggregation, especially regarding high-power consuming appliances with distinctive power draw characteristics for which satisfactory accuracy results have been reached. However, in order to distinguish devices with similar consumption traces and handle possible simultaneous state changes, other features should be taken into account [7] too, such as Reactive Power [6,18].

Other research works have investigated if further information could allow NILM systems to reach better accuracy results [6,13,19,20]. Such information can be directly measured (*i.e.*, Voltage, Current) or derived (*i.e.*, power peaks, Power Factor, Root Mean Squared voltage and current, phase differences, *etc.*) [7]. Furthermore, in several works [21–23], a Fourier series analysis has been performed to determine current harmonics, although the low sampling constraint allows for extracting only the lowest ones. These additional features have helped to identify non-linear loads with a non-sinusoidal current trace and to discriminate between loads with constant power and constant impedance [7].

In most works, data were sampled up to 1 kHz [6,19], while in [13] and [24], the proposed appliance classification approaches were using samples gathered every 1 and 2 min, respectively.

### 3.1.2. High Sampling Rate

High frequency sampling measurements have been considered in order to reach a higher detection accuracy, by also taking into account the transient-state.

In [14], the power shapes of transient events have been used as features; the authors have observed that the transient behavior of several appliances is different and thus can be used as a characterizing feature. In [25], the authors used as a feature the energy calculated during the "turning ON" transient event. High frequency collection also allows performing a deeper Fourier analysis and extracting higher harmonics as has been experimented in [26]. Zeifman and Roth [12] asserted that a set of harmonics (instead of a single one) can be used as complementary features of active and reactive power. In order to save resources and improve performance, Norford and Leeb [14] enhanced Hart's method introducing harmonics analysis using transient signals. In [27], Patel *et al.* have used the high frequency analysis of the voltage noise during the transient events.

### 3.2. Disaggregation Approaches

The NILM methods implemented so far can also be distinguished for the approach type. There are two ways to conceive the training phase of a learning method: supervised and unsupervised. Both of them have weaknesses and strengths [28].

A supervised approach makes use of labeled data in the training phase in order to allow the NILM system to detect device contributions from the aggregate consumption load [7]. Consequently, an increase in terms of both computational resource investments and human effort for the system startup phase has to be considered; however, it generally offers good accuracy results.

Starting from Hart's work, in 1992 [6], which made use of Finite State Machine (FSM), many other different supervised approaches have been proposed, as those based on k-Nearest Neighbor (k-NN) [29] and Support Vector Machine (SVM) [26,30]. Kramer *et al.* [31] have recently performed an analysis for comparing disaggregation accuracy results achieved by different classifiers such as SVM, NN and Random Forests. As it has been shown that the temporal transitions information could improve the disaggregation [12], few algorithms that could manage this combination have been investigated. For instance, Artificial Neural Networks (ANN) have been used in many works as they offer better extensibility, dynamicity and capability to incorporate device state transition information such as in [13,19,25]. Ruzzelli *et al.* [13] proposed a supervised NILM system, called RECognition of electrical Appliances and Profiling (RECAP), based on a single ZigBee sensor for energy monitoring clipped to the main electrical unit.

In an unsupervised approach, the system does not have any a priori knowledge about the devices and often requires a manual appliance labeling when the disaggregation phase has finished. In [32] the genetic k-means clustering has been used to isolate the Real Power and Reactive Power steady-states and to detect the number of the turned-ON devices. Zia *et al.* [33] propose an appliance behavior modeling approach which uses Hidden Markov Models on Real Power traces. One of the most recent and original unsupervised approaches is the one proposed by Kolter and Jaakkola [9] in 2012. This method consists in fact in modeling each appliance consumption behavior with a Hidden Markov Model and the aggregate consumption with the additive factorial version; the authors also proposed a new inference algorithm, called Additive Factorial Approximate MAP (AFAMAP) to separate appliances traces from the aggregated load data. Egarter *et al.* [34] propose an approach based on additive FHMM that introduces the use of Particle Filtering for estimating the appliance states.

Few recent projects have remarked on the need to provide the system with context information in order to both better characterize the appliance profiles and improve disaggregation performances.

In 2011, Kim *et al.* [35] extended the FHMM approach with an unsupervised disaggregation algorithm that uses appliances behavior information (*i.e.*, ON-duration, OFF-duration, dependency between appliances, *etc.*). With respect to Kim's work, our original contribution is based on the addition of environmental and statistical features such as respectively the user presence and the daily usage distribution of several appliances. In addition, Shahriar *et al.* [36] proposed a similar approach which uses temporal and sensing information but with the aim of performing an appliance classification of power traces of single or a combination of two devices. Furthermore, a private dataset has been used in both [35] and [36], thus non-comparable results have been produced; conversely, our work uses a public dataset [10], which is thus available also to other researchers. Several open data sets are available at this time: high frequency datasets such as BLUED (Building-Level fUlly-labeled dataset for Electricity Disaggregation) [37] or REDD (Reference Energy Disaggregation Dataset) [38]; low frequency data sets such as TRACEBASE [10] or ultra-low frequency as AMPds (Almanac of Minutely Power dataset) [39]. As BLUED and REDD include various features for each analyzed appliance, TRACEBASE, provides simple active power data for each monitored appliance. In [40] a detailed comparison among some of the above-mentioned public datasets has been published. The authors also provide semi-automatic labeling algorithm to help researchers in creating fully labeled energy disaggregation datasets. We chose TRACEBASE since it provides public low frequency power consumption traces of various devices gathered in real houses. Moreover, the whole data set is fully labeled and contains temporal information for each power sample.

## 4. Disaggregation Algorithm

This section describes the proposed new energy disaggregation algorithm. First, we briefly mention the principles of the state of the art approach we adopted and then we describe how we enhanced this approach to leverage context information (e.g., timing usage statistics and user presence).

### 4.1. The Probabilistic Data Model

The observation of the devices'consumption traces has underlined that most of them usually switch from a power consumption value to few others during each period of use; every trace can thus be considered as a set of transitions from a consumption level to the subsequent one. Consequently, the mean value of each power level with its associated variance can be regarded as a state.

The Hidden Markov Model (HMM) is a probabilistic learning method for time series where the information about the past is transmitted through a single discrete variable, precisely named "hidden state" [8]; in this work, the HMM represents the power consumption evolution as a sequence of states. Such Markov processes are labeled starting from the outputs; analyzing the observed state, the algorithm assess what is the most likely Markov model hidden state capable of generating the observed output. Each device, thus, has been modeled through an HMM according with the power states and the transition matrix which determines the probabilities of each state to evolve in another. HMMs have been treated as Factorial HMM as described in [8] to consider the independent utilization overlaps of each device; the observed output is thus composed as a state additive function of the different hidden states.

The single hidden Markov model, with its conditional independencies, is graphically represented in Figure 1a, where a sequence of observations $\{Y_t\}$ with $t = \{1, ..., T\}$, is modeled by a probabilistic relation with a sequence of hidden states $\{S_t\}$, and a Markov transition structure connecting the hidden states [8].

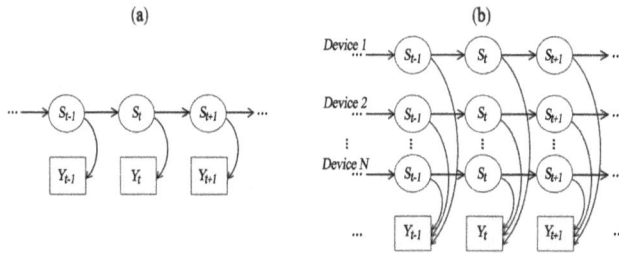

**Figure 1.** (**a**) the single Hidden Markov Model; (**b**) the additive Factorial Hidden Markov Model. Image adapted from [8]. Copyright 1997, Kluwer Academic Publishers.

The state can take one of $K$ discrete values, $S_t \in \{1, ..., K\}$, which have been extracted from the occurrence histogram composed by several power traces for each appliance, through the use of a clustering technique (*i.e.*, Gaussian Mixture Model). The transition matrix has therefore a $K \times K$ dimension and represents the state transition probabilities, $P(S_t|S_t1)$ Figure 1b shows the additive FHMM in which each independent HMM for each monitored device evolves in parallel; the sequence of observed output $\{Y_t\}$ represents the aggregate hidden states; the algorithm thus estimates which is the most probable sequence of Markov hidden states that could have produced that output.

*4.2. Inference*

As mentioned above, Kolter and Jaakkola [9] in 2012 proposed an approach based on additive Factorial Hidden Markov Model that aimed at improving inference complexity performances and avoiding local optima issues [9]. As the number of devices to disaggregate grows; in fact, the evaluation of all the possible HMM evolutions that could have generated the aggregate output, implies an increase in the computational complexity of the disaggregation process. Therefore, the authors proposed an algorithm called Additive Factorial Approximate MAP (AFAMAP) which is able to bypass the unreachable exact inference through the approximation of the Maximum *A Posteriori* Probability [9]. Kolter and Jaakkola [9] have released a Matlab version of the AFAMAP algorithm (2012); in their paper, they provide some test results and discuss the effectiveness of the algorithm compared to other inference algorithms (*i.e.*, Maximum *A Posteriori* Probability, Structured Mean Field, *etc.*) in terms of disaggregation error.

For simplicity, we do not quote the mathematical model as it is available in detail in [9] with some comparative results.

*4.3. Context-Based Disaggregation*

The contextual conditioning has been realized by adopting and extending the Conditional FHMM [35] solution, which allows integrating context information to the classical FHMM in order to obtain dynamical, rather than static, state transition matrices.

Among the various approaches made available in literature, including in particular the Conditional Random Fields [41], our choice fell on Conditional FHMM as it allowed us to easily extending the approach by Kolter and Jaakkola [9], while maintaining the use of the above-mentioned AFAMAP inference algorithm.

We selected the following types of context information:

- timing-usage statistics, which has been generated through a statistical analysis over the Tracebase dataset.

- user presence information, which has been synthetically generated for the purpose of this work. In real world cases, these data could be collected through presence sensors located in the private home rooms.

The selection of these features has been performed taking into account cost of real-life deployments. Therefore, we preferred to use a very small number of presence sensors, typically one for each room in the house, instead of a huge number of different sensors (e.g., pressure, ignition switches, movement, *etc.*).

### 4.3.1. Timing Usage Statistic Conditioning

The Tracebase dataset provides active power measurements and their relative sampling instants. This information has allowed us to evaluate the timing-usage statistics of the different devices. From the available daily measurements, the number of turning-ON events (OFF-ON transitions) of each device in time intervals of 30 min has been derived. Then, the occurrence histograms of the turning-ON events (in 24 h evaluation periods) have been generated.

The analysis of the histograms has led us to extract the information for the conditioning which is the higher or lower probability that a device has been turned ON in a specific time of the day.

For example, Figure 2a shows the occurrence histogram of a refrigerator. Relevant trends for the conditioning are not visible, due to the "always ON" nature of the device. *Vice versa*, a washing machine (Figure 2b) shows a very low turning-ON probability during night hours; this information can thus be employed to modify the state transition probabilities of this device.

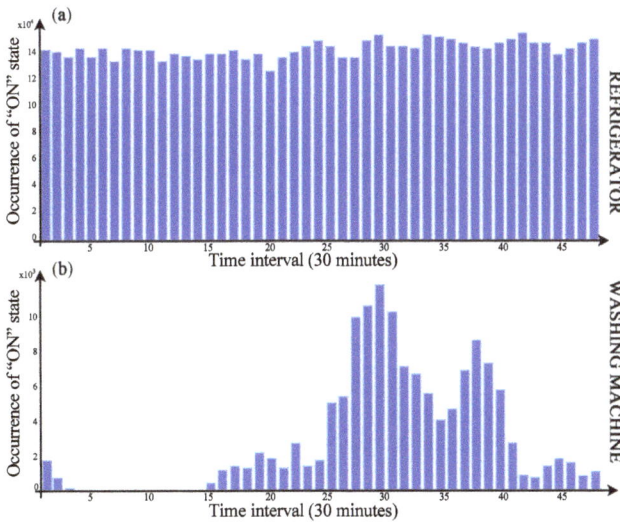

**Figure 2.** Usage statistics distribution for a refrigerator (**a**) and a washing machine (**b**).

### 4.3.2. User Presence/Absence Conditioning

The necessary information to perform this second conditioning derives from presence sensors appropriately deployed in the house. This type of conditioning consists in modifying the probability associated to the "OFF-ON" transitions of a device trace, according to the presence/absence of a user in a specific time interval. It was assumed, in fact, for specific types of appliances, the turning-ON event cannot occur without the presence of a user.

In particular, we considered two possible time intervals of observation:

- Presence/absence of users in a single time interval;
- Presence/absence of users in two consecutive time intervals.

The time interval duration is arbitrarily set as the longest consumption cycle that has been detected among the analyzed traces (*i.e.*, a washing machine cycle).

In order to clarify this critical claim, it is useful to analyze the operating characteristic of a washing machine through a three-state Markov chain: state 1 "OFF", state 2 "WASHING PROGRAM", state 3 "WATER HEATING". In Figure 3, the $P_{ij}$ terms indicate the transition probability from the state $j$ to the state $i$, typical of the single device.

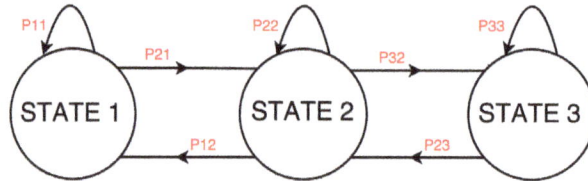

**Figure 3.** Example of a State Machine with associated transition probability modeling a washing machine operation.

The example shows that the device in examination does not provide, among the possible transitions, the transition from state 1 to state 3 without passing through state 2 ($P_{13}$ and $P_{31}$ are nil). Table 1 shows the probabilities to transit from state $i$ to state $j$ and *vice versa* in a single step.

**Table 1.** Example of a transition matrix for a washing machine.

| $\dfrac{j}{i}$ | S1 | S2 | S3 |
|---|---|---|---|
| S1 | 0.9978 | 0.0022 | 0 |
| S2 | 0.0307 | 0.9690 | 0.0003 |
| S3 | 0 | 0.0012 | 0.9988 |

User absence in the single time interval implies that:

- If the device is ON, it will continue its customary working cycle until the end of the washing cycle;
- If the device is OFF, the transition to state 2 is not possible ON during the time interval (because there are not users in its neighborhood), thus: $P_{21} = 0$ as it is shown in Figure 4 and in Table 2.

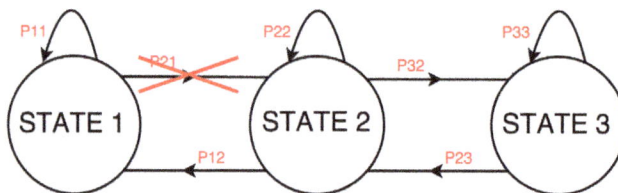

**Figure 4.** Example of a state machine for a washing machine derived by taking into account user presence information.

**Table 2.** Example of a transition matrix for a washing machine derived by taking into account user presence information.

| $\dfrac{j}{i}$ | S1 | S2 | S3 |
|---|---|---|---|
| S1 | 1 | 0 | 0 |
| S2 | 0.0307 | 0.9690 | 0.0003 |
| S3 | 0 | 0.0012 | 0.9988 |

The user absence for two consecutive time intervals, instead, implies that:

- If the device was ON before the beginning of the observation (before the first interval), it will have terminated its working cycle within the first interval; therefore, it is currently OFF (OFF in the second interval);
- If the device was OFF, it will not have had a new turning-ON, thus it is currently OFF (always in the second interval). The Markov chain for the device in the second time interval collapses in a single state, precisely the OFF state, with, as the sole possible transition, itself; the probability from state 1 to state 1 results thus unitary ($P_{11} = 1$) as shown in Figure 5 and in Table 3.

Hereafter, we will refer to these conditioning mechanisms as follows: Usage Statistic Conditioning (USC) and User Presence (UP) single/double Interval Conditioning (IC).

**Figure 5.** State machine of a washing machine taking into account user presence information for two consecutive time intervals.

**Table 3.** Transition matrix of a washing machine taking into account user presence information for two consecutive time intervals.

| $\dfrac{j}{i}$ | S1 | S2 | S3 |
|---|---|---|---|
| S1 | 1 | 0 | 0 |
| S2 | 0 | 0 | 0 |
| S3 | 0 | 0 | 0 |

## 5. Experimental Results

In this section, we describe the experimental activities carried out to validate our approach. First, we describe the dataset [10] that we used and how we extracted the HMM models for each considered appliance. We then show and discuss a meaningful disaggregation test for each context-based conditioning mechanism by providing both the graphical and the numerical disaggregation results at appliance level. Averaged disaggregation results are also discussed for four different test cases and compared them with the basic algorithm by Kolter and Jaakkola [9]. Appliance profiling has been performed using Python scripts on a machine equipped with an Intel Core 2 Duo P8400 at

2.26 GHz, 3 GB RAM; disaggregation test campaigns have been performed using Matlab version R2012a on a machine equipped with an Intel Core2 Duo CPU T7500 at 2.2 GHz, 2 GB RAM and another with Intel Core 4 i7-3610QM at 2.3 GHz, 8 GB RAM.

In order to provide experimental results which could be compared with those of other works, the precision and recall parameters [42] have been chosen. The parameters are calculated as follows:

$$Precision = \frac{True\ positive}{True\ positive + False\ positive} \tag{1}$$

$$Recall = \frac{True\ positive}{True\ positive + False\ negative} \tag{2}$$

Considering the real and the disaggregated power samples for each device:

- The true positive parameter represents the number of samples that have been correctly classified or, more precisely, the power quantity correctly assigned to that device.
- The false positive parameter represents the number of samples that have been incorrectly classified or, more precisely, the power quantity incorrectly assigned to that device.
- The false negative parameter represents the number of samples that should be but have not been classified or, more precisely, the power quantity that should have been assigned to that device but has been assigned to another or has not been assigned at all.

The *precision* parameter measures the portion of power samples that has been correctly classified among the power samples assigned to a given device. The *recall* parameter measures what power portion of a given device is correctly classified in general, also considering that samples that would belong to that device but have been wrongly assigned to another or not assigned at all.

In order to show a general parameter that could combine the results obtained through the *precision* and *recall* analysis, the *F-Measure* parameter has been considered and calculated as follows.

$$F - Measure = 2\frac{Precision\ Recall}{Precision + Recall} \tag{3}$$

Although *F-Measure* represents a statistical combination of *precision* and *recall*, in our experimentation, the first parameter has a more pertinent meaning in the single appliance disaggregation results, as it enhances the percentage of the right assigned power samples. For this reason, in our test-cases discussed below, *precision* results are shown at a single appliance level, while at a test and overall level, recall and F-Measure are also pointed out.

*5.1. Data Analysis and Pre-Processing*

As a preliminary step, we have evaluated Tracebase [10], which is the dataset that we adopted in this work, and performed few preprocessing operations on the data. Tracebase, which has been introduced in Section 3, is a public, password-protected dataset. It consists of real power consumption traces of a range of electric appliances that have been collected in more than ten households and office spaces. The trace collection script, described by Reinhardt *et al.* [10] (2012), has been configured to gather one sample per second; furthermore, every sample is stored with its timestamp. However, because of the topology of the data collection network and the encountered delays, the authors stated that traces may also show a higher or lower frequency; this physical characteristic forced us to perform an accurate data analysis and a pre-processing phase that are described below. Moreover, this dataset is conceived to perform the appliance classification, thus it provides reliable power consumption traces as they all have been detected with a dedicated smart plug. Therefore, it does not include an aggregate consumption signal. In this work, we set up a synthetic aggregate power trace consumption that is composed of a sample-sum of a selected subset of the available traces. Indeed, Tracebase includes up to 1270 monitoring traces of 122 devices of 31 different appliances types, but we used a subset made by 423 traces of 43 devices belonging to 6 types (Table 4), by selecting those devices that presented a major number of traces and less holes in the monitoring interval.

**Table 4.** Tracebase subset of appliances used in our approach. Data have been partially reproduced from [10].

| Device Type | #Appliances | #Traces |
|---|---|---|
| Coffee Maker | 5 | 39 |
| LCD TV | 10 | 94 |
| Microwave Oven | 5 | 48 |
| PC Desktop | 9 | 90 |
| Refrigerator | 7 | 130 |
| Washing Machine | 7 | 22 |

As stated by Reinhardt *et al.* [10], Tracebase presents several detection inhomogeneities; a daily power trace can, in fact, show more than one sample per second or a lack of data for some seconds. Hence, we processed all the daily power trace by normalizing each one with 86,400 samples (number of seconds in a day). We have performed the zero-padding or the average operations on the missing/surplus samples and put the obtained values in a normalized trace. These operations have become reasonable after the data analysis. For instance, when a device results in being disconnected to the electric network (OFF state), the meter has obviously gathered a zero-consuming trace; thus, we have zero-padded the missing samples and reduced in an only zero value the surplus samples gathered at the same second. Moreover, when the device is active (ON state) we have evaluated the samples immediately before and after the missing one/ones and performed an average operation of them and then filled the missing value with the obtained result. An analogous operation has been performed in the case when there was more than a sample per second.

## 5.2. Appliance Profiling

In order to extract the power levels that typically characterize an appliance consumption, we analyzed all the available traces for each appliance in our subset. As mentioned above, according to the consumption behavior and the nature of the devices, each device consumption profile can be approximately characterized by just few power states. To identify them, we generated for each type of device a power value occurrence histogram aimed at highlighting the most frequently achieved values ranges.

After this operation, a further sub-sampling operation is performed. The zero power, which corresponds to the disconnected state, could, in fact, mislead the research of accumulation values as it reasonably represents, except for the "always ON" appliances, the most frequent sample value. Therefore, a coherent sub-sampling has been applied by processing each sequence through a sliding window of fixed size (10 samples); in the case where the samples observed in the window result all 0, nine values of these will be barred from the data on which to search the state value. After this pre-processing phase, the problem of determining the intrinsic structure of the data to be grouped, in the case that only the observed values result accessible have been considered; hence the preciseness of the state extraction has been tested through clustering analysis [43].

Clustering analysis organizes the data according to an abstract structure in order to recognize groups or hierarchies of groups. A cluster is composed of a number of similar objects collected or grouped together according to a specific parameter named distance. How the distance is set up and which parameter it represents depends on the chosen algorithm and on the type of data to be processed. In our experimentation case, the objects are represented by the power values; the clustering algorithm has to evaluate and group them together in order to extrapolate few power states that could effectively describe the consumption behavior of each type of appliance. The clustering algorithm identifies few mean values and their associated variances that could represent as accurately as possible each consumption state.

To solve our problem, we made preliminary tests with some clustering algorithms; k-means [44] and Gaussian Mixture Model (GMM) [45] reported the most consistent results.

With k-means, given a set of $n$ points defined in a $d$-dimensional space $R^d$ and an integer $k$, the problem consists in determining a set of $k$ points, belonging to $R^d$, called centroids, such that each mean squared distance for each point belonging to the cluster is minimal when compared to the centroid. In our case, the centroids represent the states of the descriptive model which is associated with each device observed. This type of technique usually fails in the general categories of clustering that are based on the variance [46]. As mentioned above, we have also investigated a probabilistic approach, named Gaussian Mixture Model (GMM). It is assumed that the data are generated by a mixture of latent probability distributions in which each component represents a different group of clusters [43]. It consists in the weighted sum of $M$ components of Gaussian densities as described by the following equation:

$$p(x|\lambda) = \sum_{i=1}^{M} w_i g(x|\mu_i, \sum_i),$$ (4)

where $x$ is a $D$-dimensional data vector (e.g., measured features), $w_i$ $(i = 1, ..., M)$ represent the mixture weights and $g(x|\mu_i, \Sigma_i)$ with $(i = 1, ..., M)$ are the components of Gaussian densities. Each Gaussian component is represented by the following shape:

$$g(x|\mu_i, \sum_i) = \frac{1}{(2\pi)^{\frac{D}{2}}|\Sigma_i|^{\frac{1}{2}}} \exp\{-\frac{1}{2}(x - \mu_i)' \sum_i^{-1}(x - \mu_i)\},$$ (5)

where $\mu_i$ is the mean vector and $\Sigma_i$ is the covariance matrix. The mixture weights meet the following condition:

$$\sum_{i=1}^{M} w_i = 1.$$ (6)

There are several variants of the GMM that have just been introduced, depending on the calculation type of the parameters that describe the distribution. The choice of model configuration (number of components, dense or diagonal covariance matrices, link among parameters, *etc.*) is often determined by the amount of available data to estimate the parameters of GMM and the environment in which the GMM is applied. One of the most important attributes of GMM is its ability to form smooth approximations of arbitrary distribution densities. A GMM acts as a sort of hybrid that uses a discrete set of Gaussian functions, each with its own parameters (mean and covariance matrices), in order to permit a better modeling capability. In this paper the data model that have been associated to each device is composed from the following components:

- The data $x = \{x_1, ..., x_L\}$ represent the sample data which is in turn a realization of $X = \{X_1, ..., X_L\}$
- $X_i$ represents the $i^{th}$ data flow which is described by a $d$-dimensional feature space $\{F_1, ..., F_D\}$.
- $X$ can be divided in data that have been labeled as $X_l$ but not $X_u$.
- $K = \{k_1, ..., k_N\}$ represents the set of state classes that are associated to each device.

Therefore, our clustering problem is reduced to finding the $N$-states which better represent each monitored device [47].

We chose to adopt GMM because of its excellent characteristics of adaptability to the proposed data. This approach allows in fact to more clearly extract the device representative states characterized by an average value and its respective variance. Table 5 shows comparative results regarding state extraction obtained with the two algorithms for a refrigerator and a washing machine.

**Table 5.** Comparative results of *k*-means and Gaussian Mixture Model (GMM) clustering algorithm for extracting the power levels (mean and variance) of a refrigerator and a washing machine.

| Algorithm | Mean | Variance |
|---|---|---|
| *Refrigerator* | | |
| k-means | 66.09 | 7.42 |
| | 302.48 | 56597.83 |
| | 61.09 | 37.66 |
| GMM | 64.13 | 8.46 |
| | 491.01 | 24501.37 |
| | 30.56 | 921.85 |
| *Washing Machine* | | |
| k-means | 6.72 | 205.39 |
| | 2100.47 | 9339.21 |
| | 167.65 | 6804.05 |
| GMM | 2.00 | 0.001 |
| | 2100.47 | 9339.21 |
| | 105.28 | 2349.91 |

In the refrigerator case, *k*-means returns two mean values that are too similar (about 66 and 61) and thus results are not useful for the HMM model extraction aim. Instead, GMM reported more defined low consumption mean values together with acceptable variance values. In the washing machine case, GMM emerges for its smaller variance as the obtained mean values for each algorithm are similar.

As introduced in Section 4, the Hidden Markov Model includes the definition of a transition matrix. Therefore, we extracted the statistical model associated with each device, or, more precisely, the state transition function that models the appliance power consumption behavior with its associated probability. For each type of appliance, according to the set of states generated through the GMM, we mapped the sequence of samples in the power traces into a sequence of states. We then detected all the transition events (including the self transitions) and counted their occurrence to extract the corresponding probability. Figures 1 and 3 show the state machine and transition probability matrix that we obtained for a washing machine, respectively.

### 5.3. Context-Based Disaggregation

The experimentation has been composed of several phases. Firstly, each context conditioning has been singularly applied to the algorithm and the obtained results have been compared to those obtained in the work by Kolter and Jaakkola [9]. Secondly, disaggregation results have been evaluated considering both context information items. Each test has been performed by providing the system with the full-knowledge regarding each appliance that could compose the aggregated consumption trace (Table 4), *i.e.*, including even those turned off. As mentioned above, the aggregate consumption trace has been composed synthetically by summing the daily traces of each single appliance. In order to create the test set, we combined each daily trace of a given appliance for a given day with all the daily traces of the other appliances. First, we describe how each single context-based conditioning approach operates.

### 5.4. Usage Statistic Conditioning

In Figure 6, an aggregate power consumption trace is shown; as it can be noticed in this temporal portion, a PC-Desktop is always ON just like the Refrigerator, an LCD-TV is turned ON a little after 8:00 am and left ON until the end of the examined temporal portion. Moreover, a Coffee Maker is used in the other two daily moments.

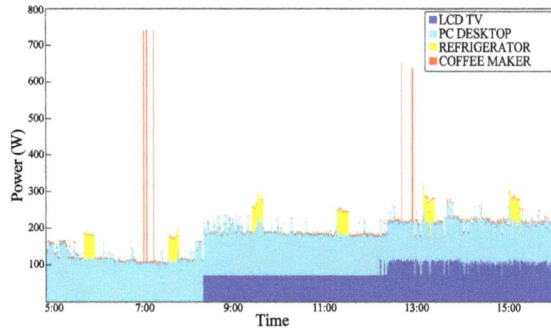

**Figure 6.** Aggregate power consumption trace—test case 1.

Figure 7 graphically shows the results obtained by applying the disaggregation algorithm by Kolter and Jaakkola [9].

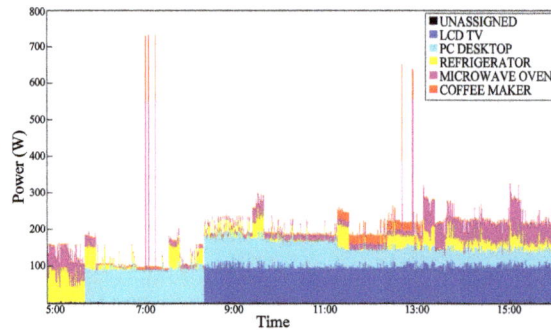

**Figure 7.** Power consumption trace disaggregated with the basic Kolter and Jaakkola [9]'s Additive Factorial Approximate MAP (AFAMAP) algorithm - test case 1.

Figure 8 shows the graphical disaggregation results obtained by applying our NILM algorithm with *Usage Statistics Conditioning (USC)*.

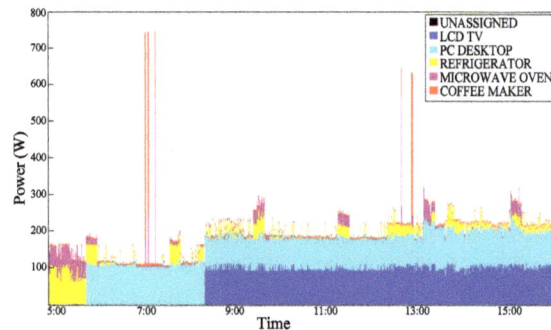

**Figure 8.** Power consumption trace disaggregated with the *Usage Statistics Conditioning (USC)*—test case 1.

An improvement can be observed; this is plausible, especially for devices that are typically switched ON and OFF in a portion of a specific time such as the coffee maker. As expected, a typical "Always ON" device such as the Refrigerator does not benefit from the effects of this type of conditioning.

**Table 6.** Precision results obtained with Kolter and Jaakkola [9]'s Additive Factorial Approximate MAP (AFAMAP) algorithm in basic and with *Usage Statistics Conditioning* version.

|  | AFAMAP [9] | USC |
|---|---|---|
| Refrigerator | 27.88% | 30.77% |
| LCD-TV | 99.28% | 100.00% |
| PC-Desktop | 50.99% | 74.07% |
| Coffee Maker | 36.14% | 77.21% |

Table 6 shows the precision obtained with the timing usage statistics conditioning compared with those obtained by using the AFAMAP algorithm [9].

*5.5. User Presence Conditioning*

The second conditioning is analyzed below. Figure 9 shows the real aggregate consumption trace of a Washing Machine with the same LCD-TV trace that has been analyzed above; the graphic shows a portion of washing cycle with a high consumption phase (corresponding to the water heating phase) in the middle. In this case, the LCD-TV disaggregation is a little worse (demonstrating that depending on the device traces combination, disaggregation precision can change) and this kind of conditioning, both in the single (Figure 10) and double (Figure 11) interval version, does not introduce relevant improvements with respect to the algorithm by Kolter and Jaakkola [9]. This is due to the fact that in this case the TV usage lasts for a very long period, probably longer than the user presence observation interval. Figures 10 and 11 shows an improvement in the Washing Machine disaggregation, obtained through the *Single Interval Conditioning* and the *Double Interval Conditioning*, respectively. The washing phase, which is characterized by low power consumption, is difficultly distinguishable; the basic algorithm, in fact, confuse it for a PC-Desktop execution (Figure 12). Although even with the *Single Interval Conditioning* few errors are encountered, a portion of washing machine consumption is well-assigned (Figure 10). The graphical results are confirmed by the precision percentage shown in Table 7.

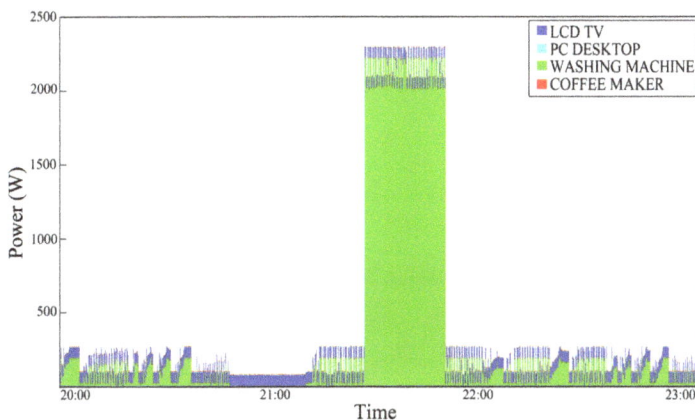

**Figure 9.** Aggregate power consumption trace—test case 2.

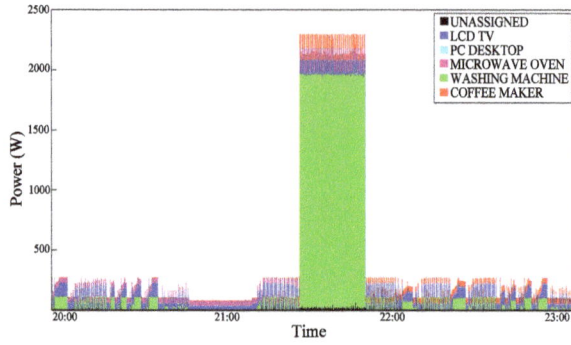

**Figure 10.** Power consumption disaggregation obtained with the *UP Single Interval Conditioning*—test case 2.

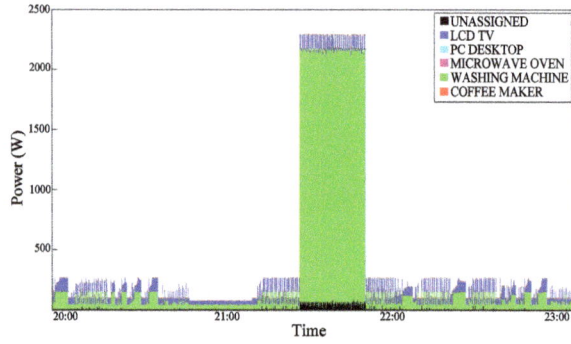

**Figure 11.** Power consumption disaggregation obtained with the *UP Double Interval Conditioning*.

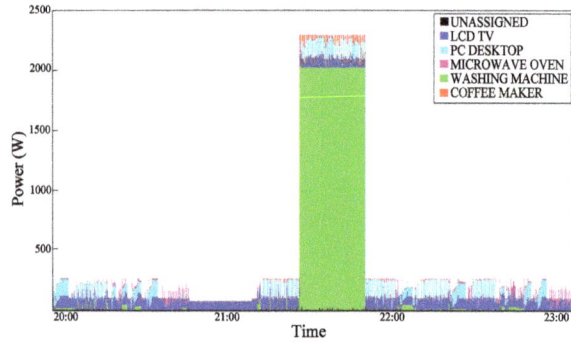

**Figure 12.** Power consumption disaggregation obtained with Kolter and Jaakkola [9]'s AFAMAP algorithm—test case 2.

**Table 7.** Precision percentages comparison between Kolter and Jaakkola [9]'s AFAMAP algorithm and the User Presence Single and Double Interval Conditioning.

|  | AFAMAP [9] | UP Single IC | UP Double IC |
|---|---|---|---|
| Refrigerator | 70.73% | 78.50% | 94.78% |
| LCD-TV | 93.86% | 90.54% | 94.04% |

*5.6. Discussion*

Table 8 compares average results of four tests using the basic Kolter and Jaakkola [9]'s algorithm, the *User Presence Single Interval Conditioning*, the *User Presence Double Interval Conditioning*, the *Usage Statistics Conditioning* and a combination of the last two conditioning mechanisms executed together. In almost all cases, a disaggregation precision average improvement is observed with respect to the basic algorithm. Even if the combination of the usage statistics conditioning with the double interval conditioning is better in most cases, percentage-wise, the most effective is the *User Presence Double Interval Conditioning*. As regards the *Recall* parameter, the average results are a little worse than the *precision* ones. This is caused by the nature of this parameter that, by definition, also considers the wrongly assigned or unassigned samples of a given device. However, the *recall* improvement over the basic algorithm tightly depends on the analyzed test case, as, for example, a greater quantity of not assigned power samples can worsen this value.

**Table 8.** Average *precision/recall* percentage results comparison among each tested algorithm for 4 test cases.

|  | AFAMAP [9] | UP Single IC | UP Double IC | USC | UP Double IC + USC |
|---|---|---|---|---|---|
| Test 1 | 44,72/73,88 | 61,23/81,11 | 61,44/74,12 | 48,68/72,98 | 61,36/73,84 |
| Test 2 | 49,63/80,49 | 43,06/69,39 | 57,48/70,46 | 57,40/73,45 | 58,18/70,68 |
| Test 3 | 44,37/81,73 | 54,83/86,97 | 69,11/71,59 | 54,18/80,74 | 68,94/71,43 |
| Test 4 | 50,76/59,51 | 41,52/70,74 | 53,27/55,26 | 53,43/55,01 | 52,92/54,38 |

The experimentation campaign, carried out with the complete test set over the basic algorithm for each conditioning, has highlighted the average improvements that have been reported in Table 9.

**Table 9.** Average *Precision* and *F-Measure* improvements over the basic AFAMAP algorithm [9].

|  | Context-Based Conditionings | Precision | F-Measure |
|---|---|---|---|
| 1 | User Presence single interval conditioning | $\approx 3\%$ | $\approx 2\%$ |
| 2 | User Presence double interval conditioning | $\approx 12\%$ | $\approx 14\%$ |
| 3 | Usage Statistics conditioning | $\approx 6\%$ | $\approx 3\%$ |
| 4 | The combination of 2) and 3) | $\approx 13\%$ | $\approx 14\%$ |

As can be observed, even though the *recall* parameter apparently worsened the disaggregation results at test level, the *F-Measure* evaluation parameter, through the whole test set, reports a significant improvement as well as the *precision* parameter.

## 6. Conclusions

In this article, we proposed a new energy disaggregation algorithm that takes into account context-related information that can be gathered from low-cost sensors and statistical analysis of energy consumption data. With respect to most existing works, which are based on the analysis of data collected at a high sampling frequency [14,25,26], our contribution consisted of investigating a disaggregation approach on energy monitoring data collected at low frequency. This choice has the following advantages: it is possible to use low-cost and widely available smart meters and data storage and transfer tasks are less resource demanding. Context features (e.g., user presence and device usage consumption patterns) have been exploited to improve the statistical model of each appliance.

Results of testing activities and their comparison with a state of the art solution are encouraging. In the future, it would be useful to extend the proposed approach to include the use of additional context information (e.g., profile of users, weather information, *etc.*) in order to improve the disaggregation algorithm as well as to enhance the proposed approach with optimization algorithms and suggestion mechanisms to help consumers in saving energy costs.

Moreover, tests described in this work are based on the use of data available from a publicly accessible dataset, *i.e.*, Tracebase [10]. We believe that the adoption of open data sets in this field may speed up research and innovation processes by favoring repeatable research and easing the comparison of different approaches.

**Acknowledgments: Acknowledgments:** The authors would like to thank Antonino Giordano and Luca Capannesi from the University of Florence for their technical support, and Pino Castrogiovanni and Fabio Bellifemine from Telecom Italia for fruitful discussion.

**Author Contributions: Author Contributions:** Francesca Paradiso contributed to the design of the proposed algorithm, carried out the experimentation activities and the analysis of results. Federica Paganelli supervised the research activities, and then contributed to the refinement of the proposed algorithm. Samuele Capobianco conceived and implemented the appliance profiling component. Francesca Paradiso and Federica Paganelli wrote the paper. Dino Giuli contributed to the advancement of the paper.

**Conflicts of Interest: Conflicts of Interest:** The authors declare no conflict of interest.

## References

1.    Energy Efficiency Status Report 2012. Available online: https://setis.ec.europa.eu/sites/default/files /reports/energy-efficiency-status-report-2012.pdf (accessed on 15 January 2016).
2.    Electric Power Annual 2008. Available online: http://large.stanford.edu/courses/2012/ph240/doshay1/ docs/034808.pdf (accessed on 15 January 2016).
3.    The effectiveness of feedback on energy consumption. Available online: http://www.globalwarmingisreal.com/ energyconsump-feedback.pdf (accessed on 15 January 2016).
4.    Ehrhardt-Martinez, K.; Donnelly, K.; Laitner, J. *Advanced Metering Initiatives and Residential Feedback Programs*; American Council for an Energy Efficient Economy: Washington, DC, USA, 2010.
5.    Mills, B.; Schleich, J. Residential energy-efficient technology adoption, energy conservation, knowledge, and attitudes: An analysis of European countries. *Energy Policy* **2012**, *49*, 616–628.
6.    Hart, G.W. Nonintrusive appliance load monitoring. *Proc. IEEE* **1992**, *80*, 1870–1891.
7.    Zoha, A.; Gluhak, A.; Imran, M.A.; Rajasegarar, S. Non-intrusive load monitoring approaches for disaggregated energy sensing: A survey. *Sensors* **2012**, *12*, 16838–16866.
8.    Ghahramani, Z.; Jordan, M.I. Factorial hidden Markov models. *Mach. Learn.* **1997**, *29*, 245–273.
9.    Kolter, J.Z.; Jaakkola, T. Approximate inference in additive factorial hmms with application to energy disaggregation. In Proceedings of the International Conference on Artificial Intelligence and Statistics, La Palma, Canary Islands, Spain, 21–23 April 2012; pp. 1472–1482.
10.   Reinhardt, A.; Burkhardt, D.; Zaheer, M.; Steinmetz, R. Electric appliance classification based on distributed high resolution current sensing. In Proceedings of the 2012 IEEE 37th Conference on Local Computer Networks Workshops (LCN Workshops), Clearwater, FL, USA, 22–25 October 2012; pp. 999–1005.
11.   Appliance Profile Specification. Available online: http://ict-aim.eu/fileadmin/user/files/deliverables/ AIM-D2-3v2-0.pdf (accessed on 19 January 2016).
12.   Zeifman, M.; Roth, K. Nonintrusive appliance load monitoring: Review and outlook. *IEEE Trans. Consum. Electron.* **2011**, *57*, 76–84.
13.   Ruzzelli, A.G.; Nicolas, C.; Schoofs, A.; O'Hare, G.M. Real-time recognition and profiling of appliances through a single electricity sensor. In Proceedings of the 2010 7th Annual IEEE Communications Society Conference on Sensor Mesh and Ad Hoc Communications and Networks (SECON), IEEE, Boston, MA, USA, 21–25 June 2010; pp. 1–9.
14.   Norford, L.K.; Leeb, S.B. Non-intrusive electrical load monitoring in commercial buildings based on steady-state and transient load-detection algorithms. *Energy Build.* **1996**, *24*, 51–64.

15. Farinaccio, L.; Zmeureanu, R. Using a pattern recognition approach to disaggregate the total electricity consumption in a house into the major end-uses. *Energy Build.* **1999**, *30*, 245–259.
16. Baranski, M.; Voss, J. Nonintrusive appliance load monitoring based on an optical sensor. In Proceedings of the 2003 IEEE Bologna Power Tech Conference Proceedings, Bologna, Italy, 23–26 June 2003, Volume 4.
17. Baranski, M.; Voss, J. Genetic algorithm for pattern detection in NIALM systems. In Proceedings of the 2004 IEEE International Conference on Systems, Man and Cybernetics, The Hague, The Netherlands, 10–13 October 2004, Volume 4, pp. 3462–3468.
18. Drenker, S.; Kader, A. Nonintrusive monitoring of electric loads. *IEEE Comput. Appl. Power* **1999**, *12*, 47–51.
19. Chang, H.H.; Chien, P.C.; Lin, L.S.; Chen, N. Feature extraction of non-intrusive load-monitoring system using genetic algorithm in smart meters. In Proceedings of the 2011 IEEE 8th International Conference on e-Business Engineering (ICEBE), Beijing, China, 19–21 October 2011; pp. 299–304.
20. Figueiredo, M.B.; de Almeida, A.; Ribeiro, B. An experimental study on electrical signature identification of non-intrusive load monitoring (NILM) systems. In *Adaptive and Natural Computing Algorithms*; Springer: Ljubljana, Slovenia, 2011; pp. 31–40.
21. Liang, J.; Ng, S.K.; Kendall, G.; Cheng, J.W. Load signature study—Part I: Basic concept, structure, and methodology. *IEEE Trans. Power Deliv.* **2010**, *25*, 551–560.
22. Laughman, C.; Lee, K.; Cox, R.; Shaw, S.; Leeb, S.; Norford, L.; Armstrong, P. Power signature analysis. *IEEE Power Energy Mag.* **2003**, *1*, 56–63.
23. Li, J.; West, S.; Platt, G. Power decomposition based on SVM regression. In Proceedings of the 2012 IEEE Proceedings of International Conference on Modelling, Identification & Control (ICMIC), Wuhan, China, 24–26 June 2012; pp. 1195–1199.
24. Paradiso, F.; Paganelli, F.; Luchetta, A.; Giuli, D.; Castrogiovanni, P. ANN-based appliance recognition from low-frequency energy monitoring data. In Proceedings of the 2013 IEEE 14th International Symposium and Workshops on a World of Wireless, Mobile and Multimedia Networks (WoWMoM), Madird, Spain, 4–7 June 2013; pp. 1–6.
25. Chang, H.H.; Lin, C.L.; Lee, J.K. Load identification in non intrusive load monitoring using steady-state and turn-on transient energy algorithms. In Proceedings of the 2010 IEEE 14th International Conference on Computer Supported Cooperative Work in Design (cscwd), Shanghai, China, 14–16 April 2010; pp. 27–32.
26. Applying support vector machines and boosting to a non-intrusive monitoring system for household electric appliances with inverters. Available online: http://citeseerx.ist.psu.edu/viewdoc/summary?doi=10.1.1.40.222 (accessed on 25 January 2016).
27. Patel, S.N.; Robertson, T.; Kientz, J.A.; Reynolds, M.S.; Abowd, G.D. *At the Flick of a Switch: Detecting and Classifying Unique Electrical Events on the Residential Power Line (Nominated for the Best Paper Award)*; Springer: Innsbruck, Austria, 2007.
28. Bishop, C.M. *Pattern Recognition and Machine Learning*; Springer: New York, NY, USA, 2006; Volume 4.
29. Larose, D.T. *k*-Nearest Neighbor Algorithm. In *Discovering Knowledge in Data: An Introduction to Data Mining*; Wiley Online Library: Hoboken, NJ, USA, 2005; pp.90–106.
30. Srinivasan, D.; Ng, W.; Liew, A. Neural-network-based signature recognition for harmonic source identification. *IEEE Trans. Power Deliv.* **2006**, *21*, 398–405.
31. Kramer, O.; Klingenberg, T.; Sonnenschein, M.; Wilken, O. Non-intrusive appliance load monitoring with bagging classifiers. *Logic J. IGPL* **2015**, doi:10.1093/jigpal/jzv016.
32. Unsupervised disaggregation of appliances using aggregated consumption data. Available online: http://users.cis.fiu.edu/ lzhen001/activities/KDD2011Program/workshops/WKS10/doc/SustKDD2.pdf (accessed on 15 January 2016).
33. Zia, T.; Bruckner, D.; Zaidi, A. A hidden markov model based procedure for identifying household electric loads. In Proceedings of the IECON 2011-37th Annual Conference on IEEE Industrial Electronics Society, Melborune, Australia, 7–10 November 2011; pp. 3218–3223.
34. Egarter, D.; Bhuvana, V.P.; Elmenreich, W. PALDi: Online load disaggregation via particle filtering. *IEEE Trans. Instrum. Meas.* **2015**, *64*, 467–477.
35. Kim, H.; Marwah, M.; Arlitt, M.F.; Lyon, G.; Han, J. *Unsupervised Disaggregation of Low Frequency Power Measurements*; SIAM: Phoenix, AZ, USA, 2011; Volume 11, pp. 747–758.

36. Shahriar, M.; Rahman, A.; Smith, D. Applying context in appliance load identification. In Proceedings of the 2013 IEEE Ninth International Conference on Natural Computation (ICNC), Shenyang, China, 23–25 July 2013; pp. 900–905.

37. Anderson, K.; Ocneanu, A.; Benitez, D.; Carlson, D.; Rowe, A.; Berges, M. BLUED: A fully labeled public dataset for event-based non-intrusive load monitoring research. In Proceedings of the 2nd KDD Workshop on Data Mining Applications in Sustainability (SustKDD), Beijing, China, 12 August 2012; pp. 1–5.

38. REDD: A public data set for energy disaggregation research. Available online: http://redd.csail.mit.edu/kolter-kddsust11.pdf (accessed on 15 January 2016).

39. Makonin, S.; Popowich, F.; Bartram, L.; Gill, B.; Bajic, I.V. AMPds: A public dataset for load disaggregation and eco-feedback research. In Proceedings of the 2013 IEEE Electrical Power & Energy Conference (EPEC), Halifax, NS, USA, 21–23 August 2013; pp. 1–6.

40. Pereira, L.; Nunes, N.J. Semi-Automatic Labeling for Public Non-Intrusive Load Monitoring Datasets. In Proceedings of the 4th IFIP/IEEE Conference on Sustainable Internet and ICT for Sustainability, Madrid, Spain, 14–15 April 2015.

41. Conditional random fields: Probabilistic models for segmenting and labeling sequence data. Available online: http://repository.upenn.edu/cgi/viewcontent.cgi?article=1162 (accessed on 19 January 2016).

42. Melamed, I.D.; Green, R.; Turian, J.P. Precision and recall of machine translation. In Proceedings of the 2003 Conference of the North American Chapter of the Association for Computational Linguistics on Human Language Technology: Companion Volume of the Proceedings of HLT-NAACL 2003–Short Papers-Volume 2, 27 May–1 June 2003; Association for Computational Linguistics: Edmonton, Canada; pp. 61–63.

43. Fraley, C.; Raftery, A.E. How many clusters? Which clustering method? Answers via model-based cluster analysis. *Comput. J.* **1998**, *41*, 578–588.

44. MacQueen, J. Some methods for classification and analysis of multivariate observations. In Proceedings of the Fifth Berkeley Symposium on Mathematical Statistics and Probability, Oakland, CA, USA, 1967; Volume 1, pp. 281–297.

45. Reynolds, D. Gaussian mixture models. In *Encyclopedia of Biometrics*; Springer: New York, NY, USA, 2009; pp. 659–663.

46. Kanungo, T.; Mount, D.M.; Netanyahu, N.S.; Piatko, C.D.; Silverman, R.; Wu, A.Y. An efficient k-means clustering algorithm: Analysis and implementation. *IEEE Trans. Pattern Anal. Mach. Intell.* **2002**, *24*, 881–892.

47. Qian, F.; Hu, G.M.; Yao, X.M. Semi-supervised internet network traffic classification using a Gaussian mixture model. *AEU-Int. J. Electron. Commun.* **2008**, *62*, 557–564.

*future internet*

MDPI

*Article*

# Senior Living Lab: An Ecological Approach to Foster Social Innovation in an Ageing Society

Leonardo Angelini [1,*], Stefano Carrino [1], Omar Abou Khaled [1], Susie Riva-Mossman [2] and Elena Mugellini [1]

[1]  HumanTech Institute, University of Applied Sciences and Arts Western Switzerland, Fribourg 1705, Switzerland; stefano.carrino@hes-so.ch (S.C.); omar.aboukhaled@hes-so.ch (O.A.K.); elena.mugellini@hes-so.ch (E.M.)

[2]  School of Nursing Sciences La Source, University of Applied Sciences Western Switzerland, Lausanne 1004, Switzerland; su.riva@bluewin.ch

*  Correspondence: leonardo.angelini@hes-so.ch; Tel.: +41-26-429-6745

Academic Editor: Dino Giuli
Received: 16 April 2016; Accepted: 29 September 2016; Published: 21 October 2016

**Abstract:** The Senior Living Lab (SLL) is a transdisciplinary research platform created by four Universities that aims at promoting ageing well at home through the co-creation of innovative products, services and practices with older adults. While most living labs for ageing well are focused on Information and Communication Technologies (ICTs), this social laboratory adopts a transdisciplinary approach, bringing together designers, economists, engineers and healthcare professionals to develop multiple forms of social innovation using participatory methods. The SLL is based on an ecological approach, connecting professionals and users in a cooperative network and involving all of the stakeholders concerned with ageing well, such as existing associations, business entities and policy-makers. Three main themes for the co-design of products and services were identified at the beginning of the SLL conception, each sustained by a major business partner: healthy nutrition to cope with frailty, improved autonomous mobility to foster independence and social communication to prevent isolation. This article shows the innovative transdisciplinary approach of the SLL and discusses the particular challenges that emerged during the first year of its creation, investigating the role of ICTs when designing products and services for older adults.

**Keywords:** living labs; older adults; ICTs

## 1. Introduction

We live in an ageing society where the increasing older population depends on the shrinking working-age population. The EUROPOP2013 study estimates that in the EU the demographic old-age dependency ratio (people aged 65 or above relative to those aged 15–64) will increase from 27.8% in 2013 to 50.1% in 2060. Therefore, the EU would move from having four working-age people for every person aged over 65 years to only two working-age people [1]. This will have an important economic impact on the healthcare and pension systems. In this context, it is crucial to promote initiatives for ageing well, supporting prevention and healthy lifestyles. The EU has supported several campaigns to promote ageing well, in particular through information and communication technology research, during the 7th Framework Programme and the Horizon 2020 (H2020) program, but also with dedicated programs, such as the Active and Assisted Living (AAL) Programme.

Making successful innovation for older adults hides several challenges that have to be addressed from the onset of the research project. For at least two decades, it has been advocated that individual behavior change is not sufficient in order to ensure an effective improvement of older adults' health [2]. An ecological action, i.e., a change also in the environment of the individual is often needed to support the promotion of healthy lifestyles and ensure their effectiveness [2]. This implies a change not only in infrastructure, but also to the existing supporting communities, policy-makers, healthcare professionals and business actors. This approach obtained good results for promoting older adults' health in France and Quebec [3]. In this context, the older adults as individuals still have an important role, since their needs should be well understood and they should be implicated, contributing throughout whole design processes of new products, services and practices. The Senior Living Lab [4,5] aims at promoting healthy ageing at home and strives to innovate for and with older adults through transdisciplinary research, creating products and services that can not only meet the functional requirements for which they have been developed, but that are also appealing, easy to use and affordable.

Exploiting an ecological approach, the Senior Living Lab (SLL) conducts community-based participatory research in the French-speaking part of Switzerland (Romandy) with a transdisciplinary team. The core of the Senior Living Lab is constituted by four academic partners with different areas of experience: business, design, engineering and nursing within the University of Applied Sciences and Arts Western Switzerland. Working together, these four domains allow the academic platform to tackle the challenges of co-creating with older adults, adhering to an ecological approach. Moreover, this academic core is reinforced by companies, associations and institutions willing to innovate in-tandem with older adults.

In this article, we aim to provide an overview of the key methodologies that are important for discussing the transdisciplinary approach and the role of technologies in a living lab for older adults, which is our main contribution. In light of this purpose, this article has the two-fold goal of: (1) presenting the transdisciplinary methodologies needed to design a living lab for older adults (highlighting the activities and interventions carried out during the first year of life of the laboratory) and the impact that they have in the functioning of the living lab; (2) discussing the role of ICTs when co-creating with older adults. Indeed, while on the one hand, the ICTs have the potential of improving the quality of life of older adults, on the other hand, older adults are often concerned by the potential negative impacts of ICTs on their lives. In this context, we propose a different solution: while most living labs for older adults conduct research and make innovation specifically in the domain of ICTs, the Senior Living Lab puts technology in the background, focusing more on general product and service innovation, using technology as an invisible support.

In the next sections, we describe the research context with respect to existing living labs, the goals of the Senior Living Lab and the participatory methodology used to co-create products, services and practices. The insights obtained during the first year open a discussion about the challenges of innovating when following an ecological approach. We think that our experiences could help other living labs or institutions that would like to co-create products and services with older adults.

## 2. Background

### 2.1. Living Labs

The living lab is a concept introduced by professor William Mitchell at the Massachusetts Institute of Technology (MIT) [6]. In its first meaning, the focus was on the observation of the living patterns of people inhabiting a smart house for several days. The basic idea that still remains in all of the current instantiations of the living lab concept is to include the users in a value-creation process.

According to Pallot et al. [6] *"Living Labs are standing at the crossroads of different society trends like citizens engaged into a more participative approach, businesses and local authorities as well as user communities are gathering within public-private–people partnership initiatives. They are also at the crossroads of different paradigms and technological streams such as* Future Internet, *Open Innovation, User co-Creation, User Content Creation and Social Interaction (Web2.0), Mass Collaboration (i.e., Wikipedia), and Cloud Computing where the Internet is the cloud, also named "the disappearing IT infrastructure"."*

Nowadays, the living lab concept is broader, and multiple definitions exist. This has brought forward the creation of different entities that coexist under the same large umbrella concept of living lab. In Europe, in order to coordinate the work, provide best practices and to disseminate the results achieved in the different living labs, an international federation of benchmarked living labs has been birthed: The European Network of Living Labs (ENoLL) [7].

According to the definition of the ENoLL [8] *"Living Labs are defined as user-centred, open innovation ecosystems based on a systematic user co-creation approach integrating research and innovation processes in real life communities and settings. In practice, Living Labs place the citizen at the centre of innovation, and have thus shown the ability to better mould the opportunities offered by new ICT concepts and solutions to the specific needs and aspirations of local contexts, cultures, and creativity potentials."*

Schuurman et al. (2013) [9] highlight the necessity of a clear conceptualization of what a living lab is. The authors assess that this is still a task in progress, but propose a four-fold categorization founded on living labs based on a literature review that empirically validated 64 ICT living labs from ENoLL. The four general living lab types that they propose are: (1) American living labs; (2) testbed-like living labs; (3) living labs focused on intense user co-creation; and (4) living labs mainly as facilitators for multi-stakeholder collaboration and knowledge sharing. According to this categorization, the Senior Living Lab has a closer relation with the third typology of living labs, in which the main goal is the co-creation of new services and products and the collection of information on the usage context with ethnographic approaches. According to Schuurman et al., *"These Living Labs can focus on the early development phases of needs analysis and (iterative) design, where, based on an identified problem, a solution is developed in close interaction with end-users.".*

We prefer this categorization to the one proposed by Leminen et al. [10] for different reasons. This last categorization proposes to differentiate living labs based on which actor drives their activities: utilizer-driven, enabler-driven, provider-driven and user-driven living labs. We believe that this latter categorization poorly fits our conception of the Senior Living Lab in which all of the actors play an important role. Certainly, the final users are at the center of the laboratory (thus, the Senior Living Lab is closer to user-driven living labs) but a consistent effort has been required to balance and take into account the impact of all of the different stakeholders.

In the co-creation process of the Senior Living Lab, we have considered the previous definitions focusing in particular on the necessity of adopting an ecological approach around the final users, i.e., the older adults. As we will explain in the next sections, this means that, differently from most of the existing living labs, the SLL does not follow a technology-centric approach, but focuses on the older adults' needs and their ecosystem.

An additional element that is often considered crucial for a living lab is to put the laboratory at the center of a Public-Private-People Partnership (4P). The foundation behind these new environments is to open academic and business boundaries to harvest creative ideas and exploitable facilities that can be found among the different project stakeholders.

Few scientific papers can be found about living labs. Schuurman et al. [11] underlined this lack, thanks to a systematic review of existing research articles in this domain. Furthermore, they concluded that many articles were merely descriptive, without proposing theoretical advances or empirical results. The lack of empirical results in scientific papers can be explained by the role that private companies may play inside a living lab. Companies can request scientific partners to limit or omit the publication of most of the results in order to avoid leaking confidential information.

Shuurman et al. [9] also highlighted that most living labs were still in an exploratory phase, which implicitly raises the problem of living labs' sustainability. Mastelic et al. [12] suggested to use the business model canvas [13] to investigate the sustainability of living labs that are members of ENoLL. Their research showed that most living labs are in an exploration phase and that only a few have been able to move to the exploitation phase. Baccarne et al. [14] suggest an open business model innovation approach, involving stakeholders in the definition of the business model itself. Conversely, Doppio and Pianesi [15] propose a service paradigm in which a host offers services to a guest inside the living lab.

In this paper, discussing the first year of life of the Senior Living Lab, we will focus on the theoretical innovations that we propose for our specific goal of designing products and services for older adults and on the controversial role of ICT for this target population.

*2.2. Living Labs and Older Adults*

Considering the specific target of the Senior Living Lab, we have identified several European living labs dealing with challenges linked to healthy ageing and older adults. For the analysis that we present in this section, we focus on living labs that are members of ENoLL, using the data publicly available about their activities at the end of 2015. From the complete list of ENoLL living labs available on the ENoLL website [7], we individuated the living labs working with older adults. The selection and classification of existing living labs was performed separately by two SLL members; complete agreement was achieved after discussion. The living labs were classified according to the following criteria analyzing the information on the ENoLL website and the respective living lab websites:

- Work on ICT (25)
  - Work only on ICT (10)
  - Home automation/domotic (10)
  - Smart objects (4)
  - Telemedicine/tele-assistance (11)
  - Robots (3)
- Have multiple target population and not only older adults (12)
- Have a specific goal (17)
  - Autonomy (11)
  - Isolation (2)
  - Specific illness (8)
  - Counselling (2)
- Work with older adults' caregivers and doctors (9)
- Have an ecological approach (8)

The number of living labs respecting each criterion are highlighted in parentheses. It is worth noting that some living labs may comply with more than one criterion (even in the same category).

Our research highlighted 28 living labs spread across Europe working in this specific field or strictly related domains (Austria 1, Belgium 6, England 2, France 7, Germany 1, Ireland 1, Italy 2, Spain 5, Sweden 2, Switzerland and Hungary 1). Some of the living labs are strictly linked together and work in complementary domains (e.g., Care living labs in Belgium) or belong to a national/regional platform linking different labs acting in different domains (e.g., InnovaPuglia, which links 79 local living labs in the Puglia region in Italy). The considered living labs are listed in Appendix A.

It is worth noting that in more than the 43% of the cases, older adults are not the only target population of the reviewed living labs. In fact, often, the proposed interventions focus also on people in the environment of the older adults (e.g., caregivers, families, doctors and nurses) or people with psycho-physical conditions that are similar to those of older adults (e.g., hospitalized people, impaired people or people with particular diseases or illness).

As mentioned in the previous paragraphs, a universally-accepted definition of living lab does not exist and, therefore, the adopted approaches are quite different and the modalities of users' involvement in the co-creation process are very heterogeneous. However, one result is particularly evident: living labs are very often linked to ICTs (in several living lab definitions, ICT is indeed a requirement [16]). In fact, most of the projects born inside European living labs aim at creating innovation thanks to novel technologies. In detail, about 90% of the living labs analyzed have projects dealing with the co-creation of ICTs, whereas the other 10% deal more with the creation of services impacting socio-economical aspects or creating innovation in medical technologies. Projects span different areas of ICT, but among the different services and solutions proposed, the development of home automation solutions (35%) and telemedicine or tele assistance services (39%) clearly stand out. In addition, most of the living labs deal with specific challenges linked to ageing, such as autonomy (39%).

In 21% of the living labs, actions are undertaken to have an impact on the older adults' ecosystem via the training of caregivers, nurses or doctors in several domains (and particularly, in the use of new technologies) or organizing informative events and workshops.

Even if all of the living labs aim at creating public-private-people partnerships, only 28% explicitly aim to enlarge this collaboration beyond the partners of the living lab, and just a few are explicitly focusing on an ecological approach.

Although the most fundamental notion behind the concept of the living lab is the integration of the users in the co-creation process, a unique definition on how the user should be involved is missing. This, again, has led to the creation of different methodologies under the umbrella concept of the living lab. For example, two typical approaches can be considered at the extremes of the living lab concept. On one side of the spectrum, we have cases in which a product already exists in a quasi-complete form and the users are only involved in the testing and fine-tuning of the product. On the other side of the spectrum, we have users actively participating from the beginning in the co-creation process, highlighting actual needs. The first approach leads to services and products based on business needs. Thanks to the support of the companies, these products and services can easily enter the market, but risk to not attend to the real needs of the target population. The second approach, on the other hand, allows the generative process to focus on concrete needs of the target population; however, such projects risk not reaching the market since the ideas proposed can be beyond the core business of the companies involved in the living lab. It is worth noting that those approaches and the risks evoked are not uncommon in living labs and can undermine their long-term sustainability. The understanding of the functioning of the living lab by all of the partners is therefore mandatory in order to prevent possible conflicts.

Differently from most of the existing living labs dealing with older adults, the Senior Living Lab aims at designing new products, services and practices and leaves ICTs in the background, focusing its activities on the co-design of innovation via an ecological approach. In this process, technologies play a supportive role, more or less influential according to the specific context.

### 2.3. Older Adults and Technology

Older adults' relationship with technology is heterogeneous, depending on age, gender, type of technology, availability of technology facilitators, price and many other factors [17,18]. Chen and Chan [17] reviewed several studies on technological acceptance in different contexts. Although many older adults have a positive attitude towards technology, they use it less than younger adults, and they show less interest in adopting new technologies. The explanation of the scarce older adults' adoption of technology can be found in Technology Adoption Models (TAM) [19] and in the more recent models Unified Theory of Acceptance of Use of Technology (UTAUT) [20] and UTAUT2 [18]. In particular, in the latest study, Venkatesh et al. [18] demonstrated that facilitating conditions (e.g., relatives supporting the learning of novel technologies) and price value are particularly important in relation to the behavioral intention to use technology pertaining to older women and that the habit has an effect on older men in later stages of experience.

Among the most relevant factors that influence the technology acceptance and adoption, Chen and Chan [17] cite the perceived usefulness, the perceived ease of use and personal characteristics, such as biophysical characteristics, sensation and perception, mobility, cognition, psychosocial characteristics and social relationships. In the context of assistive technology, Bright and Coventry [21] suggest that in order to obtain desirable devices, it is important to pay particular attention to the aesthetic design of the product, avoiding stigmatization and resemblance with medical devices, as well as to align the usefulness and usability with the desirability of the product.

Concerning the usability and accessibility of technology, in 2006, Pattison and Stedmon [22] analyzed the typical problems that older adults can have while interacting with mobile phones, suggesting design guidelines that can take into account the possible physical and cognitive impairments of older adults. Current touchscreen smartphones have improved the usability for older adults over traditional phones, but they still have much room for improvement, since older adults are often neglected as target users [23]. Touch interfaces have facilitated the access to digital information for older adults, thanks to improved learnability over traditional Window, Icon, Menu, Pointer (WIMP) interfaces [24], increasing the diffusion of Internet and other digital services among older adults. A recent report in Switzerland attested that the diffusion of Internet usage among older adults increased by 47% from 2009–2014, but still 44% of over 65 users lack Internet access. The report also shows a split attitude towards technology, equally distributed among technology enthusiasts, who are interested in new technologies and that have a good experience with them and those that are more reluctant [25].

The introduction of technology in older adults' lives can open several opportunities, increasing their quality of life and independence. According to the European Community, this could also have a huge economic impact on the whole society. Nevertheless, McLean [26] underlines that most initiatives that promote ICT for older adults do not really come from the actual expectations and needs of older adults. Most technology for older adults is designed and developed by engineers, as a potential solution to common problems, but without including many older adults in the design and development of this solution. Indeed, promoting e-inclusion and offering easy-to-use digital services and products often ignore the opinion of a small, but still considerable, part of older adults that do not accept technology in their lives. According to McLean [26], alternative solutions that do not make use of technology should be provided for these people.

A living lab that seeks innovation through a transdisciplinary and participatory approach could be able to address these questions and deal with the ethical issues that are particularly relevant in this field. Involving older adults from the very beginning of the product, service development and practice definition, starting from actual expressed and unexpressed needs and recurring to technology only when it is perceived as effectively useful, is the strategy adopted by the Senior Living Lab.

## 3. Goals of the SLL

### 3.1. Ageing Well at Home

As the rest of Europe, Switzerland is a country where the median age of the population is increasing and, therefore, ageing well and staying healthy is a social and also economic priority to improve the quality of life of older adults and, at the same time, to reduce the expenditure for long-term healthcare. The Senior Living Lab role in this context is to support ageing well at home for older adults (defined as 65+ people), but also including people who are approaching the retirement age (60+ people). The main population target of the project are older adults that are still functionally independent or that are entering a new phase of life in which they are confronting a form of frailty. Very different cultures and languages characterize Switzerland. This project focuses its activities on the French-speaking part of Switzerland. To achieve ageing well at home, the Senior Living Lab concentrates its efforts on three more specific axes, described in the next subsection.

### 3.2. Co-Creation of Products, Services and Technologies (Three Axes)

Three axes of investigation have been individuated through preliminary semi-structured interviews before the beginning of the project: nutrition, social relationships and mobility. For each research axis, the Senior Living Lab strives to co-create with older adults, generating new products, services and practices that existing business entities will introduce into the market.

The first axis, nutrition, aims at investigating the challenges that older adults are facing, from acquiring products and using services available in a supermarket, to promoting healthy nutrition.

For this axis, the Senior Living Lab cooperated with a well-known supermarket chain, which is interested in offering personalized services and products in order to create a better shopping experience for older adults.

The second axis investigated by the Senior Living Lab is facilitating communication with older adults to prevent social isolation and related risks. Indeed, several studies demonstrated that individuals with stronger social connections have generally better cognitive functions, better health [27] and, as a result, longer life [28]. For this axis, the Senior Living Lab is collaborating with a leading ICT provider in Switzerland.

Finally, a key factor for preventing older adults' dependency is ensuring autonomous displacements away from home. Mobility is also a crucial factor ensuring a network of social relationships and increasing the opportunity for social contacts [29]. For this reason, the Senior Living Lab is also collaborating with a local transportation company that wants to improve its offer of services for older adults.

### 3.3. Community Making and Link-Building

In order to reach older adults for co-creating products, services and practices, it is mandatory to build a community that is akin to support the Senior Living Lab. Looking for older adults "in the wild" could be time consuming and cost-demanding; therefore, creating a network through existing associations and institutions is fundamental.

Involving strategic actors at academic, associative and political levels in order to define and design the Senior Living Lab activities is crucial, ensuring an ecological approach that can bring tangible results into the older adults' everyday life. In this sense, it is crucial that the Living Lab is perceived in his role of mediator and facilitator for existing entities, instead of being perceived as a competitor, vying for resources within the network.

### 4. Methodology of the SLL

In [5], we have presented the methodologies used to co-construct social innovation inside the Senior Living Lab, referring to a social constructionist approach. The reader can refer to the article for a complete overview describing the Senior Living Lab methodologies. In this article, we aim to provide a framework depicting the key methodologies that have been implemented in the field, discussing the specificities mentioned in this article, i.e., the transdisciplinary approach, as well as the role of technologies in a living lab from the perspective of older adults that participated in-tandem with university researchers.

### 4.1. Ecological Approach

Since the ecological approach is crucial in the functioning of the Senior Living Lab, we discuss how we implemented this approach in this sub-section.

Ecological models have gained relevance in studies in many disciplines and fields (e.g., public health, sociology, biology, education, psychology, etc.) [30]. Several studies, such as [3,31], demonstrate the positive impact that can be obtained creating programs that act on multiple determinants and targets. In particular, inside the scientific community for several years now, there is a broad consensus that programs based on ecological approaches are more effective than traditional ones [2,30,32].

Nevertheless, existing programs for health promotion for older adults tend to adopt a traditional approach targeting individual determinants, instead of more innovative approaches that consider the target population and the related ecosystem [33]. This is mainly because programs based on ecological approaches implement multiple goals, require very different know-how and are therefore typically harder to design and realize in practice. In fact, ecological approaches need to consider multiple determinants of health and, therefore, necessitate a transdisciplinary team of researchers and partners. Ecological approaches should be able to act on: (a) the individuals; (b) their immediate local context; and (c) the larger context where they live, such as their community or social context.

A living lab design methodology naturally facilitates the implementation of an ecological approach. In particular:

- Living labs deal with the target users in their environment. In this way, it is possible to acquire knowledge that was not accessible in classical laboratory settings.
- A living lab is a public-private-people partnership. Therefore, it has by design the capability to provide multi-level actions and interventions around the target population.

Therefore, the Senior Living Lab distinguishes itself from the other living labs working in the same area for the involvement in its activities, and from the beginning, of all of the stakeholders around the problematic of healthy ageing. The involvement of the different stakeholders is strictly dependent on the political, societal and economic conditions of the Swiss society. For this reason, during the first year of operation, the Senior Living Lab conducted a survey of existing older adult stakeholders in the French-speaking part of Switzerland. The purpose of the survey was to identify potential partners, gain access to the older population and investigate project opportunities where the Senior Living Lab could offer its transdisciplinary competencies.

Each SLL founding partner, in particular the School of Nursing Sciences La Source, which is already in contact with several older adults' stakeholders, collected a list of potential entities that could become partners within the Senior Living Lab. During a Senior Living Lab meeting with all of the founding partners, lists of potential partners were collected. Entities belonging to the same categories have been regrouped, and a mind-map has been created.

The different entities identified as possible partners within the Senior Living Lab have been labelled in eight different categories:

- Associations: There are several local and national associations that offer a varied support for older adults: some of them offer specific support related to older age diseases (e.g., Alzheimer and Parkinson); other associations support local leisure and cultural activities; others defend older adults' rights or gather volunteers for older age support.
- Medical-social entities: Several public and private entities offer specific medical and social support to older adults. Among them, Pro Senectute (https://www.prosenectute.ch) and the Swiss Red Cross (http://www.ifrc.org) offer several services and social activities for older adults. Several nursing retirement homes and associations for home help and care exist. Their interests are represented at regional (cantonal) level by the respective associations.
- Medical entities: This category embraces the proper Swiss medical entities, such as local hospitals and University hospitals.
- Business entities: Every business entity that offers products and services for the population could be a potential Senior Living Lab partner.
- Urbanism: The entities that design and regulate the Swiss population's everyday environment have an important role, ensuring the environmental accessibility to older adults, encouraging older adults' mobility and preventing older adults' isolation, accidents and falls.
- Political entities: Political institutions have a critical role in ensuring older adults' well-being. Besides formal Swiss political institutions, such as cantons, communes and cities, there are other national-level institutions that defend older adults' rights from a political point of view.
- Foundations: There are few Swiss private and public foundations that support ageing-related projects.
- Academic entities: Besides the four academic institutions that founded the Senior Living Lab, many other Swiss Universities conduct research dealing with ageing well. Collaborating with the existing research centers is important because it allows the SLL to take advantage of existing knowledge, creating synergies in specific domains of application.

These eight categories include the major actors, allowing the Senior Lab to act on the aforementioned axes (immediate and larger context) typical of an ecological approach.

The relations with each of these categories have to be carefully considered in the development of a living lab. In fact, complex relations can arise slowing down the activities of a new laboratory. For instance, taking into account the "associations" category, when creating a new laboratory, it is very important to clarify as soon as possible the living lab's role in relation to the existing entities in order to avoid unjustified rivalries.

Even if this research was conducted in Switzerland, this taxonomy can be easily extended and applied to categorize living labs stakeholders in Europe and abroad.

*4.2. Community-Based Participatory Research in the SLL*

Participatory approaches are commonly adopted by most living labs and are particularly important to meet the needs of older adults and avoid misconceptions. User-centered design techniques are adopted broadly in the fields of service design and human-computer interaction. Indeed, involving the user from the beginning of each project is mandatory to ensure that the designed product or service will be effectively used or purchased by the older adults [34].

Community-Based Participatory Research (CBPR) focuses on ecological perspectives, as described in the previous section, to act on the different determinants of health. CBPR should consider the individuals in the community, their local context, as well as the larger context in which the community acts, such as their societal context [35].

The Senior Living Lab has used different user-centered techniques to investigate older adults' needs using a bottom-up strategy. Starting from the lower, individual level, in these first years of life, the Senior Living Lab has focused its activities on ethnological studies (more details about these user-centered techniques are presented in Section 5):

- Focus groups are a typical method to debate about open topics, such as nutrition, mobility and interpersonal communication, but also to investigate the perception of existing products, services and practices in relation to older adults.
- Another powerful tool to collect insights is the World Café [36], which aims at investigating specific questions through a participatory approach.
- Shadowing techniques [37] are a form of ethnographic study that allow discovering the latent needs of older adults, by following them during their daily activities and listing observations.

Organizations and entities working at the local and national levels have been contacted in order to establish partnerships to design ecological interventions and actions.

The insights collected through the different ethnographic studies are then incorporated into the next phase of creative thinking, generating innovation in the specific field addressed by the project in order to collectively imagine a response to older adults' most pertinent needs.

*4.3. Products, Services and Practices Design: A Transdisciplinary Approach*

The previous section described how the Senior Living Lab collects the user requirements through a participatory approach. In the design thinking methodology [38], this is often called the "empathize" phase, which follows the "define" phase in which all of the observations are collected, refined and discussed to understand the user requirements. The next phase, "ideate" involves a generative process for collecting as many ideas as possible that could solve the problem or that could rethink completely the way a certain service or product is used. The next step, "prototype" is materializing one or more of these ideas into a mock-up or prototype that the user could test to provide feedback. Indeed, "test" is the last phase of the design thinking innovation process, which iterates the different phases until an acceptable working solution is found.

Having the end-user at disposal during the whole process for assessing the ideas and testing the different prototypes is crucial in obtaining usable and appealing products and services. Meanwhile, it is also important to highlight that the whole process is conducted by a transdisciplinary team, able to conceive of multi-facetted ideas thanks to the diverse backgrounds of the participants within the creative process. In a transdisciplinary team, each member can contribute with their own knowledge and expertise, but collective efforts determine the most suitable ideas or approaches. In fact, this co-creative scenario, in which older adults, companies and institutions participate and exchange know-how, facilities and needs, is particularly apt for the design of interventions acting on multiple determinants. The concept of a transdisciplinary team is slightly different from an interdisciplinary team, which is typically intended to have members working on their own solutions separately and, during the final steps, merging together the different responses towards a final solution [39].

The presence of healthcare professionals and engineers within the team ensure that functional requirements are met, while design professionals ensure that the product and services are usable, innovative and appealing for the user. At the same time, business professionals ensure that the product and service are marketable, generating value for the involved business entities. Combining these multiple perspectives with the collaboration of the final users allows a more balanced impact of technologies, i.e., making them closer and more acceptable to the user, while at the same time, bringing forth a product or service that is valuable in terms of the market.

*4.4. Business-Oriented Sustainability*

The Senior Living Lab was born in 2014 as an academic research project founded by a Swiss private foundation. One of the objectives of the project is transforming the Senior Living Lab into an autonomous entity that is able to operate sustainably over a long period of time. In order to achieve this goal, the Senior Living Lab should offer a catalogue of unique services that business entities and other institutions would pay to hire.

The Senior Living Lab's purpose it to establish a sustainable functioning process to continue innovation for older adults after the end of the initial research project. While in the first phase the initial funding will support the development of a product or service for each of the three axes described in Section 3.2, after the end of the funded period, the Senior Living Lab should be able to operate autonomously. As in other living labs [15], the business model should be based on a catalogue of services that could be offered to strategic partners, which can be found among the list of older adults' stakeholders, presented in Section 4.1.

A typical service offered by existing living labs is the access to the end-users for testing and co-designing services and products. Following this idea, since business entities often struggle to find end-users to co-design and test products, the Senior Living Lab could act as a mediator or facilitator between the older adults, the business entities and the other stakeholders that would like to involve this particular category of users in their product or service definition.

Besides this typical service, the Senior Living Lab could offer unique services that other associations are not able to propose, benefiting from the internal competences of the transdisciplinary team and the strong links with the associative, institutional and political worlds. Thanks to the background of the academic partners, the Senior Living Lab could provide unique insights and ideas exploiting participatory design techniques. Indeed, the individual competencies of each founding university provide knowledgeability for designing products or services that align functional and health-related requirements with usability and aesthetics [19], ensuring at the same time the economic viability of the proposed product or service.

In order to analyze the sustainability of the Senior Living Lab and clearly identify a strategy for the next years, we have applied the Business Model Canvas (BMC) [13] to project the functioning of the laboratory in a long-term view, as suggested by the work of Mastelic et al. [12]. The solutions considered are strictly linked to the functioning of the University of Applied Sciences in Western Switzerland, and a deep focus on this analysis goes beyond the scope of this article. However, it is worth highlighting that, as a result of the BMC analysis, a living lab working as a facilitator among the different stakeholders was the more interesting approach, while the University partners should have the role of financing the bare minimum, guaranteeing the functioning of the laboratory solely in the case that private or third party funds are not available. This is particularly interesting since it comports a pivot in the role of the Senior Living Lab. According to the taxonomy of Schuurman [9], the role of the living lab would move from a (3) living lab focused on intense user co-creation to (4) a living lab acting mainly as a facilitator for multi-stakeholder collaboration and knowledge sharing. This shift makes perfect sense if we consider that the first years of life a living lab strives to understand the concrete needs of the users, develop methodologies and, finally, create the relations with the different communities of stakeholders. In this scenario, the living lab is no longer incarnated just by the university's research entities, but becomes an autonomous entity in which the academic members will become (privileged) stakeholders of the laboratory.

## 5. Senior Living Lab Activities

This section highlights the activities performed by the Senior Living Lab during its first year of life. In particular, we focus on the description of the types of participatory activities that we have conducted and the role played by the users in those activities. During the first year, our primary focus was on nutrition and mobility, two of the three main axes. The work in relation to the communication axis has been planned during a two-year period.

The activities carried out were the following:

- Two world cafés (one per axis)
- Two focus groups (one per axis)
- Six ethnographic studies: shadowing (three per the axis nutrition; three per the axis mobility)
- Three workshops (one per axis)

These activities are summarized in Figure 1. Whilst older adults participated in all of the activities, industrial partners were involved in only specific events. In addition to the previous activities, a documentary film was made with the participation of the senior women's group in Valais. A round table discussion was organized at the Avant-Premiere film presentation, highlighting hopeful aging themes expressed within the documentary. This cultural axis brought together community stakeholders in a dialogical space devoted to envisioning an ecological approach to citizen-centered approaches to healthy aging.

Since these are ongoing activities, an analysis of the results cannot be published yet (results are to be considered confidential with the involved economical stakeholder), and therefore, this section focuses on the methodologies used during the different activities.

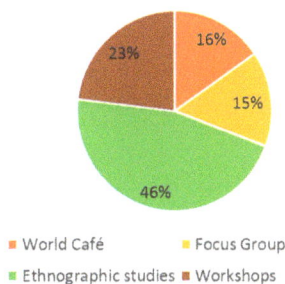

**Figure 1.** Activities of the SLL (Senior Living Lab) during the first year.

### 5.1. Case Study 1: Improving the Supermarket Experience

The first case study was conducted in collaboration with a well-known Swiss supermarket chain. Within the axis of nutrition, the aim of the project is improving the senior-friendliness of their supermarkets, in particular in the cantons of Valais and Vaud.

### 5.1.1. Activities

In order to better understand the unexpressed needs of the older adults concerning the shopping experience and the theme of nutrition, different activities have been carried out, which include 3 ethnographic studies using the shadowing technique, 1 focus group and 1 world café. The three ethnographic studies consisted of following two women living alone and a married couple during a shopping experience, not only inside the supermarket, but also before and after the shopping experience, at home and during the displacement from home to the supermarket. The ethnographic studies have been conducted in different cities and allowed the Senior Living Lab to understand the older adults' shopping and nutrition habits.

Moreover, ethnographic observations and a focus group including supermarket chain employees were carried out in Valais, with a "senior" group constituted from the regional Senior Living Lab participants, this needs assessment phase provided the methodological foundation for data collection, storying the shopping experience, a key factor within the topic of nutrition.

The second participatory activity was a world café that involved 26 participants. Four questions were discussed at each table: the first question investigated the most valuable elements identified in relation to the older adults' healthy aging, imagining innovative ways to obtain them; the second one investigated the important things that they would like to find in the supermarket chain; the third one investigated the interpersonal exchanges that developed in the supermarket chain; the fourth question explored novel ideas to foster ageing well together.

### 5.2. Case Study 2. Mobility with Public Transportation

The second case study was conducted in collaboration with a Swiss provider of public transportation. Similar to the previous case study, the aim of the different activities was to analyze and improve the senior-friendliness of public transportation. For this analysis, our geographical focus was the Canton of Vaud.

### 5.2.1. Activities

As with the previous topic, different activities have been carried out, including 3 ethnographic studies using the shadowing technique, 1 focus group and 1 World Café.

The ethnographic studies consisted of following three different older adults with different physical capabilities and limitations travelling with public transportation in the Lausanne area. The study was not limited to investigating the activities inside the means of transport, but the whole travel experience: buying tickets, waiting for the transportation means, getting on, travelling, etc.

The focus group was structured to analyze different issues linked to mobility, beginning with a more generic perspective and ending with specific transportation service issues for older adults living in the Lausanne area. The focus group was structured in four parts in order to investigate four different elements. The first part of the focus group aimed at investigating challenges and issues related to mobility from a general perspective. The second part investigated challenges and issues related to public transportation. The third part confronted the older adults with the specific solutions provided nowadays by the current services and that are specifically designed for older adults. The fourth part was dedicated to analyzing specific communication strategies and products designed by the transportation service provider.

### 6. Discussion: Role of ICT in the SLL and Participatory Design Practices

The activities conducted during the first year allowed the researchers within the Senior Living Lab, in tandem with the "senior researchers", to collect many specific insights concerning the two mobility and nutrition axes. However, we consider these insights out of the scope of this paper. Instead, in this section, we will focus on and discuss the insights obtained about the SLL methodology. In particular, we want to highlight the difficult role of ICTs when some older adults involved in the co-creation process are generally unsympathetic towards those more technical solutions.

## 6.1. Technology Fear versus Innovation

Information and communication technology have received increasing attention from the research community to support ageing well. The H2020 and the AAL programs particularly encourage the application of ICT as a support for older adults' everyday life. While there are clear benefits of using technologies for this purpose, older adults are often afraid of technology. Even if this statement is controversial (some studies stated that older adults have positive attitudes toward technologies [40]), our experience confirmed this attitude during the different focus groups that we organized. In particular, the reason for this fear is not only related to the difficulty to learn a new interface, but it is often related to the fear of losing social contacts with other people because of the technology. Unfortunately, this fear is often well justified, because many digital services tend to replace the human operator, in order to cut costs and speed up the operation. While this is acceptable for younger people, older adults, especially those living alone, often regret that in our increasingly digitalized society, they are losing many opportunities for social exchanges with other people.

As a solution toward this attitude, the Senior Living Lab and other living labs for ageing well should not be presented to older adults as a technology mediator. In addition, products and services that would be co-designed with older adults should not necessarily involve technology. Nevertheless, technology plays a relevant role in innovation and, if correctly designed, has the potential of having a huge impact on the quality of life of elders. This is a real contradiction, and the fear of new technologies can block the development and the creation of innovations potentially helpful to older adults. This is particularly relevant in participatory design approaches in which older adults are involved from the beginning of the conceptual design of a solution and can therefore ostracize the adoption of novel technologies.

As the results of the first year's activities (world cafés, focus groups, ethnographic studies and workshops involving more than 100 participants), we have learned that:

(1) Invisible technologies (technologies that are transparent to the user's eyes as in the Mark Weiser definition [41]) can reduce the fear of novel technologies. This kind of solution is well accepted by older adults in particular if they address a specific need. Invisible technologies are an important field of study in the human-computer interaction field. For instance, in one of the focus groups organized by the Senior Living Lab, older adults had shared consensus in disliking solutions based on the generic idea of creating objects enriched by displays or touch screens. However, they appreciated the idea of having "augmented" physical objects, if they are already confident with the utilization of the object. For instance, the participants liked the idea of having a technologically-enhanced magnifying glass to use in a supermarket context to retrieve additional information about a product by just using the glass on the product. In this example, older adults do not have to learn how to use a new tool or navigate in a touch screen menu, they already know how to use it. They just have the possibility to use it with new objects to obtain different information. This is a clear example of invisible technology.

(2) The use of technology is highly stratified by age, gender, marital status and educational background. This is confirmed also from other studies, such as [39]. Therefore, designers of interfaces have to carefully consider these aspects and adapt their solutions to the target population for which they are designing a technological solution. In other terms, in most of the cases, "older adults" is not a good enough segmentation for the design of ICT-based solutions.

(3) The usefulness of a technology is not enough to make a product or service acceptable: it is crucial to avoid the older adults' stigmatization. Products or services that are explicitly presented as being designed for older adults are likely avoided by seniors, since it is considered equivalent to the admission of belonging to this group. In addition, a clear category of "older adults" does not exist, since age is not always the pertinent criteria allowing researchers to segment the aforesaid population. Instead, physical and mental disabilities together with the cultural baggage of the person play a major role, impacting user needs.

## 6.2. Addressing Older Adult's Needs versus Stigmatization

Designing products and services with older adults specifically thought through with their lens and perception of needs leads to products and services that are often labelled as such and that have only older adults as target customers. In this case, older adults tend often to consider these products and services as stigmatizing, especially if used in front of their relatives and friends. Products and services designed with older adults to address their specific needs should be designed and sold as a product or service for all and not as specifically designed for older adults. This is sometimes counterintuitive for both the business entities and for the Senior Living Lab. Indeed, the business entities would like to improve their products and services for older adults and want to advertise those efforts. Similarly, the Senior Living Lab is interested in labelling the designed products and services with its logo as an advertisement of its work and also as a guarantee for the older adults of a "senior friendly" product or service. At the time of writing this article, it remains to be known if a Senior Living Lab "senior friendly" label would be accepted by the older adults or if it would be perceived as stigmatizing.

Concerning this point, the lesson learned from the first year of co-design with older adults is that designers should be able to present to the older adults a concept in the real scenario of utilization as soon as possible in order to ensure the acceptability of the product.

## 6.3. Usefulness versus Usability

Technology can provide many functionalities for older adults, addressing different needs with different features. However, adding several functionalities to the same product or service can make the interface more complex, thus more difficult to understand and use. This is a typical issue of general-purpose commercial devices, which often cannot satisfy older adults' needs because of usability issues. Conversely, devices designed specifically for older adults are often oversimplified and suffer from stigmatization. From our previous experiences, we noticed that tangible interaction could help hiding the complexity of technology behind a physical interface that exploits the previous knowledge of the user through meaningful interaction metaphors. In particular, the seamless integration of technology into everyday objects could help older adults accept it.

## 6.4. Participatory Approach versus Older Adults' Expectations

Beyond technologies, a key challenge during the creation of the Senior Living Lab has been its wish to be founded via a participatory approach. The goal of this approach was to ensure the functioning of the laboratory around the specific needs of the local target population having the older adults participating within the co-creation process of the laboratory. During the kick-off event and during a first world café, we asked older adults what they would expect from the Senior Living Lab, with a particular focus on the three action axes described in Section 3.2. The results of this participatory kick-off that sought to collaboratively define the Senior Living Lab's main actions were themselves controversial. Indeed, many attendees at the kick-off expected a concrete and clear action plan from the Senior Living Lab, aiming as soon as possible to obtain tangible results. In fact, at this time, the older adults did not understand their role and how they could belong to the laboratory even though an introductory presentation tried to explain their important role and the contributions that they could make as senior participants.

The lesson learned from this experience is that, in the first period of life of a living lab, it is important to balance the participatory approach with more concrete actions. These actions should be established and discussed internally by the Senior Living Lab's constituting members, since the initial participatory process could be perceived as a lack of internal organization that could slow-up the results of the project from the older adults' perspective. Participatory governance must be jointly practiced and learned. Most participants are not familiar with grass-root processes that engender collaborative results.

*6.5. Participatory Approach versus Business-Oriented Projects*

One of the hardest challenges of the Senior Living Lab was combining the participatory approach with the business-oriented purpose of the Senior Living Lab. Although the development of business-oriented services and products together with business actors is fundamental to bring the innovation to the end-user, the collaboration with those actors was sometimes perceived negatively by the older adults. In other terms, while older adults are keen to collaborate with academic institutions, the same cannot be said for business entities. In a few occasions, older adults expressed their concern of being exploited by the companies and that the Senior Living Lab was financed by business entities to collect information about them and, in the end, just to sell more products and services.

With the goal of mitigating this effect, we proposed an advisory board composed of older adults that would be involved in the decision-making process of the laboratory. Older adults deeply appreciated this idea since they felt more "in control" and not just passive actors in the laboratory. The lesson learned is that it is very important to illustrate correctly to the older adults their active role in a living lab, i.e., that they are part of the laboratory and not just testers. At the same time, it is also important to underline the benefits that they and their peers will receive in the long term via their participation (e.g., products that respond to their concrete needs). In fact, it should be clear for the participants that they are actively contributing to the development of a product that will improve their quality of life. This clarification is important for all of the stakeholders.

*6.6. Ethics and Related Concerns*

Working with a vulnerable population can give rise to ethical issues. This is particularly true in connection with medical programs needing to acquire medical-level data. Thanks to the different figures involved in the laboratory, and in particularly to health specialists, these issues are handled taking advantage of their experience and, in addition, an ethical commission will evaluate and approve all of the interventions that deal with medical and personal information.

It is therefore important to highlight that if no personal or medical data are acquired and stored and as the interventions do not impact on the medical dimensions of senior lifestyles, the ethical commission is not required to be involved.

The initial participants were invited to the Senior Living Lab kick-off meeting, contacting associations and institutions that daily collaborate with older adults and also via the personal network of the Senior Living Lab participants. For logistical and organizational reasons, we limited the number of participants to about 30 people (including researchers, partners from the industry and older adults). During this meeting, older adults interested in participating in the work of the laboratory provided contact information.

The organization of the following activities has been based on these very positive first results, and this has engendered the solidification of a contact network facilitating the organization of activities.

Finally, to meet older adults' privacy concerns, all of the sensible data acquired during the interventions are anonymized and hosted in private university servers.

*6.7. Participatory Approach versus Concrete Needs*

Understanding and taking into account the end-users' needs is mandatory in order to apply an open innovation approach and to design products and services that respond to such needs. However, tackling the question of older adults' individual needs could be a challenging task on many occasions, because older adults' needs are heterogeneous and can vary according to the personal health status, the different physical and cognitive disabilities, as well as the personal social and cultural backgrounds. Older adults often expose very particular needs that could not be addressed in each project, especially if a business-oriented goal is set to ensure the operationalization of the innovation action. While ensuring accessibility to products and services to minorities is an ethical and often legal duty for the researchers and the public business operators, taking into account the minor needs of all of the users is practically impossible and economically not sustainable in most of the cases.

In this context, during the first phases of co-creation, it is crucial to individuate the older adults' shared needs, without ignoring obstacles that could hinder the access to products and services to people with particular cognitive or physical impairments. Sometimes, accessibility problems could be solved with slight design changes, which can also improve the overall usability of the product or service for other groups of people. When accessibility problems could not be solved without increasing the complexity and the cost of the product or service, it is probably preferable to provide alternative solutions and services that can better meet the needs of the minority. In this case, a human mediator could be the most economic and appreciated solution. Shared value is a central vision that ultimately federates the interests of elders, stakeholders and communities. The configuration of a mediating platform is a process emerging from participatory governance, giving value to citizen-centered solutions to healthy aging.

## 7. Conclusions and Perspectives

In this paper, we have presented the Senior Living Lab, a living lab created in Switzerland and aiming at promoting ageing well at home pathways through the co-creation with older adults of innovative products, services and practices. The Senior Living Lab is based on participatory design and on an ecological approach.

During the first year of life of the Senior Living Lab, many insights have been gathered. This paper presents the insights relative to the SLL methodology and the role of the ICT in this context.

Compared to the existing living labs within the ENoLL network, the Senior Living Lab proposes a different approach with regard to novel technologies. In particular, we have seen that older adults are generally insecure when dealing with ICTs. The causes can be different and vary with the older adult's age, culture and health conditions, as well as with the application context. A major issue is the association of technologies with a diminution of human and social contacts. From our perspective, the primary objective of a living lab is not to design novel ICT solutions, but to provide ecological solutions to specific issues. In this scenario, technologies are fundamental, but can be made invisible or acceptable to the final users if carefully designed involving the user in the process from the very beginning. In fact, we have observed that when applying participatory design methodologies, older adults will rarely propose solutions involving novel technologies. On the other hand, they are open to discuss new technology propositions and include them if they feel that the technology will not negatively affect their lifestyle (e.g., reduction of social contacts, stigmatization, etc.).

Living labs have to maintain a precarious balance between academics, companies, associations, political entities and the final users, while working together in an ecological approach. During the first year of life, the Senior Living Lab has experimented with the necessity of have a transdisciplinary team in order to face the very different challenges that this rich heterogeneity can potentially generate. Transdisciplinarity is therefore the key to enable a living lab to design and realize actual ecological interventions.

In the next year, the academic partners in collaboration with the companies will concretize the results of the first phases of co-design. Obviously, this second phase will also include the users in the implementation and evaluation of the tasks. Meanwhile, the SLL partners are working to define a business plan allowing the living laboratory to be sustainable and keep living in a long-term scenario.

**Acknowledgments:** This research project was funded by Gebert Rüf Stiftung (GRS/067/14) and the University of Applied Sciences Western Switzerland. We thank our partners Henk Verloo, Delphine Roulet Schwab, Nataly Viens Python, Luc Bergeron and Nathalie Nyffeler for their participation in this research project. We also thank University of Applied Sciences Western Switzerland School of Nursing La Source, Lausanne, Switzerland, the University of Applied Sciences Western Switzerland, Ecole Cantonale d'Art de Lausanne (ECAL), Lausanne, Switzerland, the University of Applied Sciences Western Switzerland School of Business and Engineering, Yverdon, Switzerland, and the University of Applied Sciences Western Switzerland School of Engineering and Architecture, Fribourg, Switzerland, for their contribution.

**Author Contributions:** All of the authors participated in conceiving of and designing the Senior Living Lab methodologies and experiments. Leonardo Angelini and Stefano Carrino wrote the paper, while Omar Abou Khaled, Susie Riva-Mossman, and Elena Mugellini revised it and contributed to its improvement.

**Conflicts of Interest:** The authors declare no conflict of interest.

## Abbreviations

The following abbreviations are used in this manuscript:

| | |
|---|---|
| AAL | Active and Assisted Living |
| CBPR | Community-Based Participatory Research |
| ENoLL | European Network of Living Labs |
| H2020 | Horizon 2020 |
| ICT | Information and Communication Technology |
| SLL | Senior Living Lab |

## Appendix A

Following is the list of living labs considered in Section 2.3, Older Adults and Technology (when available, a link to the living lab description is provided).

- AAL Living Lab Schwechat: http://www.ceit.at/ceit-raltec/aal-living-lab-schwechat
- Living Lab ActivAgeing (LL2A): www.activageing.fr
- Autonom' lab: http://www.autonom-lab.com/
- Living and Care Lab (LiCaLab): www.LiCaLab.com
- CASALA Living Lab: www.casala.ie
- Cyber Care Clinique Living Lab (CCCL)
- e-Care Lab: Innovative Care Living Lab In Rhône-Alpes: http://www.i-carecluster.org/presentation-living-lab/
- eHealthMadrid Living Lab
- eHealth Living Lab à Grenade
- FZI Living Lab: http://aal.fzi.de
- LUSAGE Living Lab: http://www.lusage.org/
- InnovAGE: http://www.innovage.be
- Innovate Dementia Transnational Living Lab
- Smart House Living Lab: www.lst.tfo.upm.es
- Seniorlab
- Lab4Living: www.Lab4Living.org.uk
- Silver Normandie HUB: http://miriade-innovation.fr/silver-normandie/normandiehub/
- Softex (Swedish Open Facility for Technology in Elderly Care)
- Stockholm Living Lab
- THAT: Tele Health Aging Territory
- TELESAL
- VisAge Living Lab
- AIPA: Ageing in Place Aalst
- ONLINE neighborhoods: http://openlivinglabs.eu/livinglab/online-neighbourhoods
- Careville Limburg: Moving Care: http://www.zorgproeftuinen.be/en
- Active Caring Neighborhood: http://www.zorgproeftuinen.be/en
- Apulian Living Lab on "Healthy, Active & Assisted Living" (INNOVAALab): http://www.openlivinglabs.eu/livinglab/apulian-ict-living-lab
- Living Lab Social in real environments: Ageing Lab

## References

1. European Commission. *The 2015 Ageing Report: Underlying Assumptions and Projection Methodologies*; European Union: Brussels, Belgium, 2014; pp. 1725–3217.
2. Stokols, D. Establishing and maintaining healthy environments: Toward a social ecology of health promotion. *Am. Psychol.* **1992**, *47*, 6–22. [CrossRef] [PubMed]
3. Richard, L.; Barthélémy, L.; Tremblay, M.; Pin, S.; Gauvin, L. Interventions de prévention et promotion de la santé pour les aînés: Modèle écologique. In *Guide D'aide à L'action Francoquébécois. Saint-Denis: Inpes, Coll. Santé en Action*; INPES: Saint Denis, France, 2013. (In French)
4. Angelini, L.; Carrino, F.; Carrino, S.; Caon, M.; Khaled, O.A.; Baumgartner, J.; Sonderegger, A.; Lalanne, D.; Mugellini, E. Gesturing on the steering wheel: A user-elicited taxonomy. In Proceedings of the 6th International Conference on Automotive User Interfaces and Interactive Vehicular Applications—AutomotiveUI'14, Seattle, WA, USA, 17–19 September 2014; pp. 1–8.
5. Riva-Mossman, S.; Kampel, T.; Cohen, C.; Verloo, H. The senior living lab: An example of nursing leadership. *Clin. Interv. Aging* **2016**, *11*, 255–263. [PubMed]
6. Pallot, M.; Trousse, B.; Senach, B.; Scapin, D. Living Lab Research Landscape: From User Centred Design and User Experience towards User Cocreation. In *First European Summer School "Living Labs"*; HAL: Paris, France, 2010.
7. Ashbrook, D.; Lyons, K.; Starner, T. An investigation into round touchscreen wristwatch interaction. In Proceedings of the 10th International Conference on Human Computer Interaction with Mobile Devices and Services—MobileHCI'08, Amsterdam, The Netherlands, 2–5 September 2008; pp. 311–314.
8. Atia, A.; Takahashi, S.; Tanaka, J. Smart gesture sticker: Smart hand gestures profiles for daily objects interaction. In Proceedings of the 2010 IEEE/ACIS 9th International Conference on Computer and Information Science, Yamagata, Japan, 18–20 August 2010; pp. 482–487.
9. Schuurman, D.; Mahr, D.; De Marez, L.; Ballon, P. A fourfold typology of Living Labs: An empirical investigation amongst the enoll community. In Proceedings of the 2013 International Conference on Engineering, Technology and Innovation (ICE) & IEEE International Technology Management Conferenceon, The Hague, The Netherlands, 24–26 June 2013; pp. 1–11.
10. Leminen, S.; Westerlund, M.; Nyström, A.G. Living labs as open-innovation networks. *Technol. Innov. Manag. Rev.* **2012**, *2*, 6–11.
11. Schuurman, D.; De Marez, L.; Ballon, P. Living labs: A systematic literature review. In *Open Living Lab Days*; European Network of Living Labs: Istanbul, Turkey, 2015.
12. Mastelic, J.; Sahakian, M.; Bonazzi, R. How to keep a living lab alive? *info* **2015**, *17*, 12–25. [CrossRef]
13. Osterwalder, A.; Pigneur, Y. *Business Model Generation: A Handbook for Visionaries, Game Changers, and Challengers*; John Wiley & Sons: Hoboken, NJ, USA, 2013.
14. Baccarne, B.; Schuurman, D.; Seys, C. Living labs as a navigation system for innovative business models in the music industry. In Proceedings of the ISPIM Conference on International Society for Professional Innovation Management, Helsinki, Finland, 16–19 June 2013; p. 1.
15. Doppio, N.; Pianesi, F. Co-creating an open working model for experience & living labs willing to provide services to external players. In Proceedings of the 4th EnoLL Summer School, Manchester, UK, 27–30 August 2013.
16. Eriksson, M.; Niitamo, V.P.; Kulkki, S. *State-of-the-Art in Utilizing Living Labs Approach to User-Centric ICT Innovation—A European Approach*; Center for Distance-spanning Technology, Lulea University of Technology Sweden: Lulea, Sweden, 2005.
17. Chen, K.; Chan, A. A review of technology acceptance by older adults. *Gerontechnology* **2011**, *10*, 1–12. [CrossRef]
18. Venkatesh, V.; Thong, J.Y.; Xu, X. Consumer acceptance and use of information technology: Extending the unified theory of acceptance and use of technology. *MIS Q.* **2012**, *36*, 157–178.
19. Davis, F.D.; Bagozzi, R.P.; Warshaw, P.R. User acceptance of computer technology: A comparison of two theoretical models. *Manag. Sci.* **1989**, *35*, 982–1003. [CrossRef]
20. Venkatesh, V.; Morris, M.G.; Davis, G.B.; Davis, F.D. User acceptance of information technology: Toward a unified view. *MIS Q.* **2003**, 425–478.

21. Bright, A.K.; Coventry, L. Assistive technology for older adults: Psychological and socio-emotional design requirements. In Proceedings of the 6th International Conference on Pervasive Technologies Related to Assistive Environments, Island of Rhodes, Greece, 29–31 May 2013; p. 9.

22. Pattison, M.; Stedmon, A.W. Inclusive design and human factors: Designing mobile phones for older users. *PsychNology J.* **2006**, *4*, 267–284.

23. Page, T. Touchscreen mobile devices and older adults: A usability study. *Int. J. Hum. Factors Ergon.* **2014**, *3*, 65–85. [CrossRef]

24. Holzinger, A. Finger instead of mouse: Touch screens as a means of enhancing universal access. In *Universal Access Theoretical Perspectives, Practice, and Experience*; Springer: Berlin, Germany, 2002; pp. 387–397.

25. Seifert, A.; Schelling, H.R. *Digital Seniors*; Pro Senectute: Zurich, Switzerland, 2015; p. 102.

26. McLean, A. Ethical frontiers of ICT and older users: Cultural, pragmatic and ethical issues. *Ethics Inf. Technol.* **2011**, *13*, 313–326. [CrossRef]

27. Dimatteo, M.R.; Giordani, P.J.; Lepper, H.S.; Croghan, T.W. Patient adherence and medical treatment outcomes: A meta-analysis. *Med. Care* **2002**, *40*, 794–811. [CrossRef] [PubMed]

28. Berkman, L.F.; Syme, S.L. Social networks, host resistance, and mortality: A nine-year follow-up study of alameda county residents. *Am. J. Epidemiol.* **1979**, *109*, 186–204. [PubMed]

29. Mollenkopf, H. *Enhancing Mobility in Later Life: Personal Coping, Environmental Resources and Technical Support; the Out-of-Home Mobility of Older Adults in Urban and Rural Regions of Five European Countries*; Ios Press: Amsterdam, The Netherlands, 2005; Volume 17.

30. Green, L.W.; Richard, L.; Potvin, L. Ecological foundations of health promotion. *Am. J. Health Promot.* **1996**, *10*, 270–281. [CrossRef] [PubMed]

31. Richard, L.; Gauvin, L.; Raine, K. Ecological models revisited: Their uses and evolution in health promotion over two decades. *Annu. Rev. Public Health* **2011**, *32*, 307–326. [CrossRef] [PubMed]

32. McLeroy, K.R.; Bibeau, D.; Steckler, A.; Glanz, K. An ecological perspective on health promotion programs. *Health Educ. Behav.* **1988**, *15*, 351–377. [CrossRef]

33. Richard, L.; Gauvin, L.; Gosselin, C.; Ducharme, F.; Sapinski, J.P.; Trudel, M. Integrating the ecological approach in health promotion for older adults: A survey of programs aimed at elder abuse prevention, falls prevention, and appropriate medication use. *Int. J. Public Health* **2008**, *53*, 46–56. [CrossRef] [PubMed]

34. Wilkinson, C.R.; De Angeli, A. Applying user centred and participatory design approaches to commercial product development. *Des. Stud.* **2014**, *35*, 614–631. [CrossRef]

35. Parker, E.A.; Robins, T.; Israel, B.; Brakefield-Caldwell, W.; Edgren, K.; Wilkins, D. *Methods in Community-Based Participatory Research for Health*; Jossey-Bass: San Francisco, CA, USA, 2005.

36. Brown, J. *The World Café: Shaping Our Futures through Conversations that Matter (Large Print 16pt)*; ReadHowYouWant: Culver City, CA, USA, 2010.

37. Czarniawska-Joerges, B. *Shadowing: And Other Techniques for Doing Fieldwork in Modern Societies*; Copenhagen Business School Press DK: Copenaghen, Denmark, 2007.

38. Brown, T. Design thinking. *Harv. Bus. Rev.* **2008**, *86*, 84–92. [PubMed]

39. Max-Neef, M.A. Foundations of transdisciplinarity. *Ecol. Econ.* **2005**, *53*, 5–16. [CrossRef]

40. Mitzner, T.L.; Boron, J.B.; Fausset, C.B.; Adams, A.E.; Charness, N.; Czaja, S.J.; Dijkstra, K.; Fisk, A.D.; Rogers, W.A.; Sharit, J. Older adults talk technology: Technology usage and attitudes. *Comput. Hum. Behav.* **2010**, *26*, 1710–1721. [CrossRef] [PubMed]

41. Weiser, M. The computer for the 21st century. *Sci. Am.* **1991**, *265*, 94–104. [CrossRef]

*future internet*

MDPI

*Article*

# Data-Enabled Design for Social Change: Two Case Studies

f

**Patrizia Marti [1,2,*], Carl Megens [2,3] and Caroline Hummels [2]**

1   Department of Social, Political and Cognitive Science, University of Siena, Siena 53100, Italy
2   Department of Industrial Design, Eindhoven University of Technology,
    Eindhoven 5600 MB, The Netherlands; c.j.p.g.megens@tue.nl (C.M.); c.c.m.hummels@tue.nl (C.H.)
3   School of Sport Studies Eindhoven, Fontys University of Applied Sciences,
    Eindhoven 5612 MA, The Netherlands
*   Correspondence: patrizia.marti@unisi.it; Tel.: +39-057-723-4743

Academic Editor: Emilio Ferrara
Received: 11 April 2016; Accepted: 8 September 2016; Published: 23 September 2016

**Abstract:** Smartness in contemporary society implies the use of massive data to improve the experience of people with connected services and products. The use of big data to collect information about people's behaviours opens a new concept of "user-centred design" where users are remotely monitored, observed and profiled. In this paradigm, users are considered as sources of information and their participation in the design process is limited to a role of data generators. There is a need to identify methodologies that actively involve people and communities at the core of ecosystems of interconnected products and services. Our contribution to designing for social innovation in ecosystems relies on developing new methods and approaches to transform data-driven design using a participatory and co-creative data-enabled design approach. To this end, we present one of the methods we have developed to design "smart" systems called Experiential Design Landscapes (EDL), and two sample projects, Social Stairs and [Y]our Perspective. Social Stairs faces the topic of behaviour change mediated by sensing technologies. [Y]our Perspective is a social platform to sustain processes of deliberative democracy. Both projects exemplify our approach to data-enabled design as a social proactive participatory design approach.

**Keywords:** data-enabled design; participatory design; co-creation; smart systems; behaviour change; deliberative democracy

---

## 1. Introduction

A series of pervasive, spontaneous phenomena emerge every day in connection with the use of "smart technologies". Hienz [1] states that the way we understand and embrace the data movement at present will shape how it impacts all of our futures, our lives, economies, societies, and the choices we make. This author also believes that changes will be for the better.

However, the connected and ubiquitous nature of these systems, associated with the enormous amount of data generated by billions of people using them, at an incredibly fast pace, poses new challenges in the design of such systems and confronts us with societal challenges that require a novel way of thinking about innovation.

Gardien et al. [2] refer to this evolution as a process of systemic innovation in an ecosystem approach. By ecosystems they mean "interconnected products, services and solutions that grow and adapt with the user to bring new value and meaning".

The Paradigm Framework developed by Brand and Rocchi [3] effectively describes the change in people's concept of value and meaning in ecosystems. The framework depicts a fundamental shift from the industrial and experience paradigm (from delivering a single product to a specific user, to delivering targeted experiences for customer segmentations) to the so-called knowledge paradigm, connected to delivering meaningful experiences in ecosystems, whereby both consumers and experts have become the developers of the knowledge needed to create these meaningful experiences.

The knowledge paradigm implies a change in the development of technology; it requires sensitivity with respect to how people behave and make sense of their lives as well as adaptivity to changes. New kinds of pervasive sensor-based and embedded technologies used in ecosystems demand a very different understanding from traditional user-interface design activities. People need to make sense of the composition of different system elements at the logical level (what can be done, what can go together with what and for what purpose), the functional level (how to use it), the physical level (it must be possible to see what fits together and how to manipulate it), and at the experience level (what does this mean for me, what are the implications in the long term, and does the system respect my privacy and rights?).

Proactive social participation is essential to coping with this complexity. The design of products, services and technological systems is, in fact, inextricably and inevitably linked with society, and has very profound social consequences.

Our contribution to designing for social innovation in ecosystems relies on developing and experimenting with methodologies that bring people and communities at the core of ecosystems of interconnected products and services.

In what follows we will first review related works using massive data to drive the design process. Later we define our approach to data-enabled design in ecosystems by presenting the Experiential Design Landscapes (EDL) methodology, and two sample projects, Social Stairs and [Y]our Perspective.

In the conclusion, we will discuss the challenges and potential of data-enabled design in smart ecosystems.

## 2. Related Work

Data has attracted the enthusiasm and attention of researchers for its potential in several areas of applications. A new generation of products, systems and services aims at changing people's behaviour, following and creating trends of quantified self [4], wearables [5] and Internet of Things (IoT) [6]. These new elements are always connected and present in our everyday lives and rely on increasingly more intelligent interactions to make our lives better. This introduces the realm of Big Data to design, as these new product systems and services continuously generate data. Big Data also offers new opportunities for learning to designers, as it is possible to investigate and analyse large body of information about people, their habits, preferences and behaviours, and gain insights for new designs. Marr [7] describes the 5 V's of Big Data, being Volume, Velocity, Variety, Veracity and Value. In design we see an emphasis on Variety, the diversity of data, and Value. Designers have to figure out new processes of research. They must learn how to deal with data, how to visualise them and make sense of them. They have to experiment with new approaches to observe people's behaviour remotely and to envision new forms of participation for users who they will likely never meet, in order to create new value through their designs and the underlying data.

There are different challenges in what Bogers et al. [8] call data-enabled design. They aim at developing a data-enabled design framework for designing intelligent products, services and ecosystems targeting behaviour change. They see three necessary building blocks for developing such a framework: user involvement, methodologies for technology-mediated data collection from the field, and the use of interactive prototypes [8]. Their work builds on the EDL approach we describe in this paper (see paragraph 3). Here, we take a closer look at these three challenges.

The first challenge relates to the involvement of users in the design process. Marti and Bannon [9] examine some of the difficulties one may encounter in performing user-centred design. They argue that while a user-centred perspective is required at all times in the design team, the forms of participation of users in the design process need to fit the context in the broadest sense (physical, sociocultural, emotional), and can vary significantly from that presented as the prototypical user-centred design approach.

Apparently, the turn to big data is opening up a new era in user-centred design, where people are systematically and remotely monitored, observed and profiled. To give an example, some researchers have taken advantage of the spread of mobile telephones to involve ordinary citizens as potential data gatherers, to collect every day a whole series of data accurately, all over the world, at a very low cost. An illustration of this approach can be found in the Citizen Science project of Paulos et al. [10,11], whereby mobile phones are used as 'networked mobile personal measurement instruments' e.g., to measure the quality of San Francisco air by deploying sensing systems (CO, NOx, $O_3$, temperature, humidity data and GPS) on over half the fleet of street sweepers and collecting several months of data. The various field studies presented by Paulos et al [11], using different mobile devices as measurement instruments, show that even if there are no real-time sharing mechanisms or informed interface for people involved in the study, an amazing interest of citizens in making sense of data clearly emerges. Most participants reported that from the study they learned that there were unsafe levels of air quality in their city beyond what the government or news agency had reported to them before. Many of them complained about politicians for not informing them adequately and for not enacting a legislation to improve the air quality in the name of public health. The project also stimulated participants in discussing and interpreting the data they collected every day. They also shared strategies adopted when they had captured dangerously high readings, such as finding alternate routes to be less polluted.

Other examples of new recent forms of participation in large amount of data gathering are known as Participatory Sensing (PS). This is a movement that observes and collects masses of highly detailed information that can then be used to improve quality of life in a great variety of different sectors, from environment to health and culture [12]. This approach involves citizens in the monitoring and control of the environment they live in, in the broadest sense of the term, meaning both territory and sociocultural space.

Illustrious examples of PS include the Citizen Science project described above, but also embrace those developed by the University of California, Los Angeles, where students were involved in projects ranging from improvement of transportation to recycling, water monitoring, safety and health [13]. However PS sees citizens primarily as "data gatherers" rather than partners of a sense-making and interpretation process of the generated data. The Timestreams Platform, developed by Jesse Blum et al. [14], tries to move more towards sense-making. This platform uses ubiquitous and pervasive computing for gathering and externalising a variety of data (e.g., $CO_2$, temperature, humidity, images and videos) by artists, to boost a societal discussion about climate change. The Brazilian artist Ali created a record whose speed when played was controlled by the $CO_2$ data collected on the Timestreams platform. Although citizens have a different role in this project, the gatherers of information are not the same persons as the ones interpreting the data.

So it seems that citizens can take up various roles in user-centred data-related design. As Sanders and Stappers [15] show in their map of UCD research (see Figure 1), most of the approaches to user-involvement can be mapped on a two-dimensional chart ranging from "users as subject and source of information" (user-centred design) to "users as participants and/or design partner" (participatory design, co-design, co-creation). The approaches share a fundamental issue in defining a suitable role for the users to deal with complex smart ecosystems.

An example of data-related user-centered design approach that is positioned on the left side of the figure is Cell Phone Intervention for You (CITY), where young adults are involved in user tests during which they receive behavioural interventions on their mobile phones to stimulate them to lose weight [16]. The above-mentioned examples of PS and citizen science are more centred around the middle or moving to the right when users have a more equal and participatory role (Figure 1). In our own data-enabled design approach we also try to explore the right side of this figure.

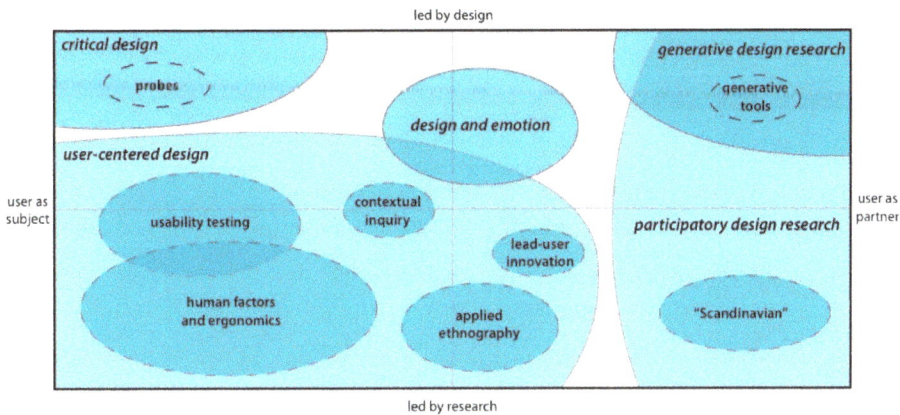

**Figure 1.** The current landscape of user-centered design research by Sanders and Stappers [15].

The second challenge is related to techniques and methodologies for performing data-enabled design research. These methodologies range from screen-based approaches such as Virtual Ethnography [17] that adapts traditional observational methods to online fieldwork to the use of sensing technology worn by the users (see PS described above), and to approaches combining opinion-based and behaviour-based data [8,18].

This also creates challenges for design researchers. As described above, the roles of users are changing due to technological developments such as ubiquitous wireless networking, an explosive growth of Web 2.0 sites and connected IoT technology. That means that the amount of data generated by citizens as "amateur professionals" (instead of the previous passive consumer who was merely digesting professionally created content) is exploding and by far exceeding the quantity of professionally created content [19]. Just imagine how much data the 600 million daily active users of Facebook, the 359 million active users of Google+ or the 300 million users of Twitter [20,21] could produce if they were involved in data-enabled design. On the one hand, technological developments such as machine learning can support the analytics of huge amounts of data to deal with the volume, variety and veracity in an automated and learning way. On the other hand, the users also play an important role in the data-enabled design process. In our opinion, the role of citizens should not merely focus on PS (participatory sensing), but on participatory sense-making, a term coined by de Jaegher and di Paolo [22] referring to the shared meaning-making process of people grounded in ongoing embodied and situated interactions in a shared action space. The data-enabled design process follows the translation from data to valuable information.

The third challenge is related to the use of interactive prototypes as a vehicle for experience and value probing, and a means of obtaining data of various kinds in an iterative way, starting early on in the design process. Prototypes can be used in many ways and forms, from the abstract to the sensorial. Prototyping creates insight and knowledge through the mechanism of reflection in and on action [23,24] and by offering a variety of experienceable prototypes through quick iterations. Designers enable people to have access to and express meaning in their everyday context, and through behavior-based data inform the designers about potential directions in which to go.

In line with these approaches, we see a value in data-enabled methods, since they offer the possibility to collect rich datasets of user's behaviours, in particular when combined with participatory approaches where people can actually contribute to the design process.

In what follows we will describe the data-enabled design method EDL, which we have developed to design "smart" systems [25,26]. The description of the method is accompanied by the presentation of two sample projects, Social Stairs [27] and [Y]our Perspective that will highlight challenges and opportunities of data-enabled design.

The projects explore data-enabled design from two different perspectives.

In Social Stairs we collected multimodal data 24/7 on people's changing or emergent behaviour in the use of stairs in a public building. The dataset was later used in participatory meetings with employees to envision social activities implying physical movement in the workplace.

[Y]our Perspective takes a different approach to data-enabled design. The social platform stimulates people in expressing a value proposition related to the place they cross in a specific moment. The system sends notifications related to the ongoing public debate and the place, and people can respond to this with their point of view. All comments are geo-localised and shared in the community, including the municipality. People can set new pools as well. Data are interpreted and discussed in public Open Space Technology meetings [28].

## 3. Experiential Design Landscapes

The Experiential Design Landscapes (EDL) method is a design research method aimed at designing for and with people in their natural environment, to find ways to support them in structurally changing their behaviour on a local scale, and to address global societal issues in the long run. EDLs are environments, be it physical or virtual, that are part of society (e.g., designated areas in cities, sports parks, virtual platforms, etc.) in which a design research team meets people in their everyday lives. EDLs can be seen as a specific type of Living Lab or Field Lab. For the exact difference, please refer to the paper by Peeters et al. [26]. The EDL method is society-aware, design-enabled and inspired. The EDL method is based on 4 process steps of envisioning, designing interventions, acquiring data, and analysing and validating this data [25] (see Figure 2).

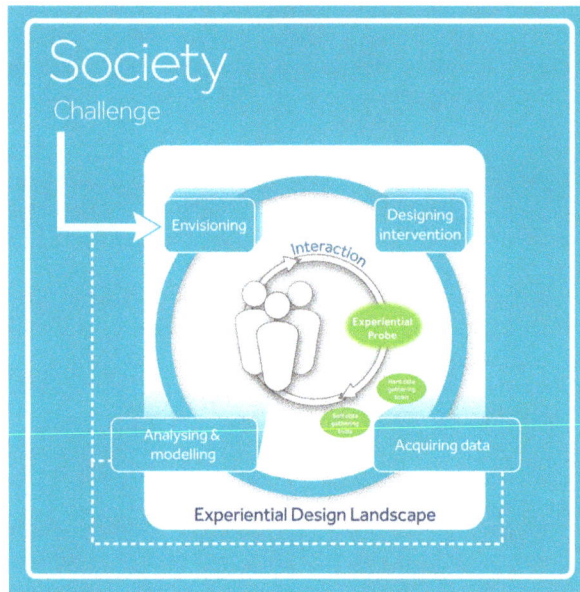

**Figure 2.** A schematic overview of the Experiential Design Landscapes (EDL) method, showing the cyclic nature of the method.

Moreover, the entire development process shifts via several stages from first explorations to value propositions that can be upscaled to a market-ready product or system, based on the growth plan as proposed by Ross et al. [29] through phases of incubation, nursery, adoption and product launch (Figure 3).

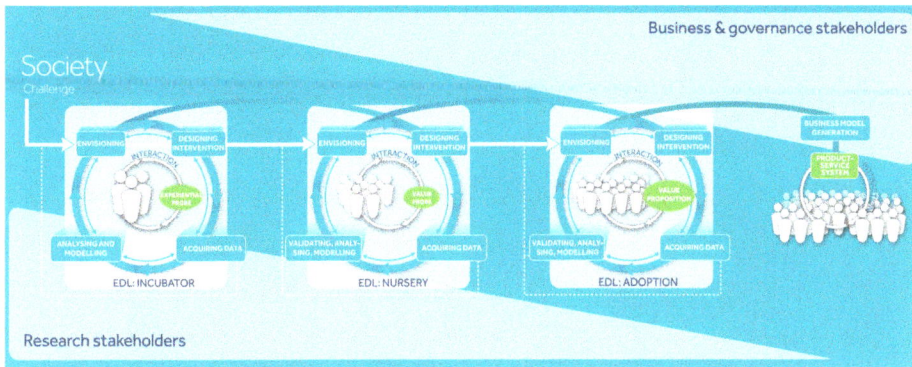

**Figure 3.** A schematic overview of the EDL method, showing the cyclic nature of the method and the progress from explorations towards value propositions.

Open, disruptive, and intelligent propositions—which we call Experiential probes (EP)—can easily be created; introduced and tailored in the EDL. These EPs are open, sensor-enhanced, networked products-service systems that enable citizens to develop new and emerging behaviour. In a parallel manner they enable detailed analysis of the emerging data patterns by researchers and designers as a source of inspiration for the development of future systems; products and services. The environments and designed propositions are instrumented with smart sensors solutions to analyse changing behaviour and new emerging patterns. Through analysis, insight is gained on the behaviour of people and the influence of the design on the interaction and on society. EP can best be described as a physicalisation of "open scripts and intentionality" that aims to play an active role in the relationship between humans and their world [30]. They are multi-stable [31]; i.e., context and relationship dependent. Depending on the relationship people have or build up with these artefacts or the context in which they are used, the probes can have different interpretations; intentionalities and identities.

The multistable interpretations, intentionalities and identities of the EP are not a result found after product launch, but can be explored, steered, challenged and radically changed during the design research process, based on people's behaviour and the design research team's vision of its future society. This creates a new process of design in which the design research team creates a dialogue through design with people in the EDL.

### 3.1. Social Stairs

Due to several causes, society is moving toward a sedentary and inactive lifestyle, with resulting consequences for the health and well-being of its citizens [32]. One of these causes is the way our daily work has changed from physical to cognitive tasks, mostly performed in a stationary state. Work occupies a large part of our day and thus is of great influence on our lifestyle and health. Van der Ploeg et al. [33] showed that sitting for more than eleven hours has such an effect on our health that exercising will not redress the damage caused by sitting this long. With the abovementioned in mind, we describe the first EDL, called Social Stairs, a design proposition for an intelligent staircase, installed in both the main buildings of the Eindhoven University of Technology and KPN, a Dutch landline and mobile telecommunications company.

On two flights of stairs, each stair detects footsteps and triggers sound accordingly. The stairs make sounds when a person walks up or down (and thus interacts with) the stairs. These interactions allow the design research team to offer a new stair-walking experience and to explore opportunities to get people using the stairs more frequently and differently. The Social Stairs aims at disrupting the usual way of transit between floors in the building. Over the course of time, different EP, i.e., various types of stair-walking experiences and accompanying soundscape designs, were developed based on this inspiration and vision of walking and playing in the Social Stairs EDL. These EP contained multiple layers of complexity and offered different sound experiences ranging from musical and rhythmical sounds to surprising sounds, echoes and even complete compositions. People could jointly explore and deepen the interaction with the Social Stairs (Figure 4). Based on the behaviour and interaction of the people on/with the Social Stairs, each EP would react differently.

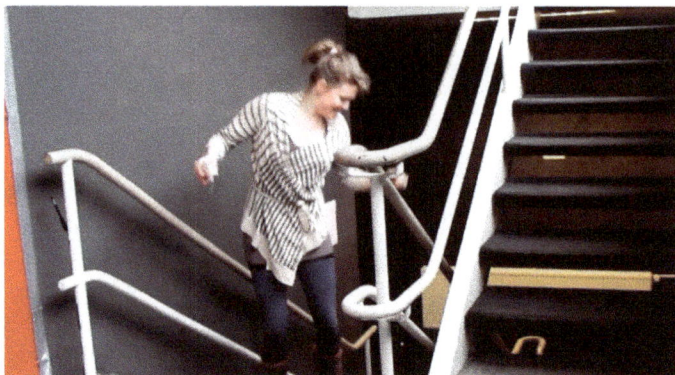

**Figure 4.** Social Stairs elicited various types of emerging behaviour. Jumping and dancing on the stairs to create music was one of the most common ones.

The Social Stairs EDL has been installed and carried out in the two separate locations where both EDLs ran for more than 6 months (Figure 5). The initial version of the Social Stairs was a result of an educational master course, where students created the first explorative probe of an instrumented and interactive stairs in the Eindhoven University of Technology main building. The first quick and low fidelity probe consisted of applying bubble wrap to a staircase. The observed resulting emerging behaviour led to new cycles of the EDL, where each stair was provided with a sensor. Two iterations have been installed between the first and second floor, on one side of the staircase, giving people the option between the Social Stairs and a normal staircase. The second iteration was built for a duration of six months, during which 4 main different EP were developed, tested, experienced and iterated. The probes differed for the sounds they created, ranging from "bloomy" echoing soundscapes to staccato drum sounds. The design research team found the different probes to result in different kinds of emerging behaviour [13]. This EDL in the main building lead to the interest of KPN to also have a Social Stairs EDL in their building, between the ground and first floor. Here a new and more robust Social Stairs was built, allowing the design research team to explore the Social Stairs concept in this new context and together with a business stakeholder.

The aim of the Social Stairs EDL was to explore ways to address the societal issue of increasing daily inactivity by being placed in an office environment. The designers wanted to find out whether they could get people working in an office environment to become more active, by both taking the stairs instead of the elevator, and increasing activity on the stairs. Although this will only lead to brief moments of physical activity during the day, it can show that designing such rich interactive environments in our sedentary environments can help to address the issue. For the designers to see whether people would change their behaviour and in what way, the Social Stairs EDL has been instrumented with sensors and cameras, to keep track of the behaviour patterns that would occur. The resulting data is again of great interest for the designers, as insights into possible emerging behaviour can further influence the design and design process towards the intended and envisioned goals.

**Figure 5.** From the first exploration to a sustainable EDL that can last for 6 months, which allowed the design research team to explore several Experiential Probes (EP) and the resulting emerging behaviour.

When introducing the EP, the design team subsequently observes, interviews and questions the people in the EDL. This way, the design research team acquired data that helped in finding or addressing specific examples of behaviour and learning more about why people express this behaviour. By means of the EDL architecture the design research team could acquire multimodal data 24/7 on people's possible changing or emergent behaviour on the Social Stairs. By saving the event logs of the steps, the design research team could keep track of which steps were taken (or not) by people and check if there were any interesting deviating behavioural patterns. Simultaneously, through the concealed miniature camera, the design research team could regularly check or real-time tap into the video to see what was going on in the Social Stairs. To gain a deeper understanding of why people were behaving in certain ways, interviews were held on the stairs as well (Figure 6). Aside from these interviews, doing maintenance and iterations on the stairs also proved to be a valuable way of gaining more insight, as people on the stairs approached the design research to share their experiences and thoughts, resulting in a more natural and spur-of-the moment user participation. On the basis of these insights the design research team could take decisions on how to iterate the current EP, or come up with a new probe.

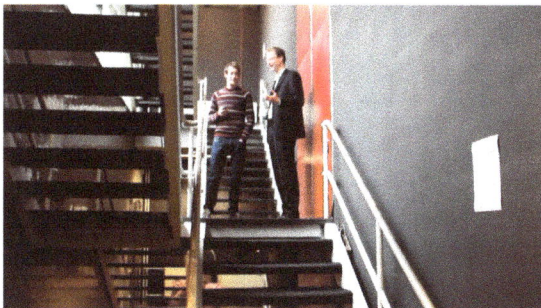

**Figure 6.** Over the 6 months of the Social Stairs EDL several interviews were held in the EDL with the aim of gathering insights into why people behaved in the ways they did.

As an example, through some early experiential probing it was found that people would engage and invite each other to the stairs. This emergent behaviour was mirrored to the design research team's initial vision on how to design for a more active and healthy society. Subsequently, this inspired them and sparked new ideas on designing different EP using louder and orchestral sounds that played with the social behaviour. Doing so would create new invitations and possibilities for behaviour on the stairs. The design research team did not predict the behaviour of people inviting other people to the stairs and playing with them together, but it did fit their vision even better on how to get people active.

Both the Social Stairs EDLs ran for over 6 months, resulting in a large dataset of recorded steps and timestamps. Through several iterations of data processing and visualisations a format was developed which allowed the design research team to see individual behaviour patterns in the data. By plotting steps over time in sequential rows the difference could be seen between "normal" stairs usage and new deviant behaviour patterns, showing new kinds of behaviour that non-instrumented stairs would not elicit. Normal stair usage would mean going up and down the stairs, hitting each (or every second) step in sequence, as can be seen in the top pattern in Figure 7. With the Social Stairs EDL new behaviour patterns emerged where people would jump, skip and repeat steps instead of normal descending or ascending behaviour. An example of this can be seen in the bottom pattern in Figure 7. Through plotting the data in this manner, the design research team was able to see human behaviour in detail, enabling them to appreciate the differences between normal and deviant behaviour patterns.

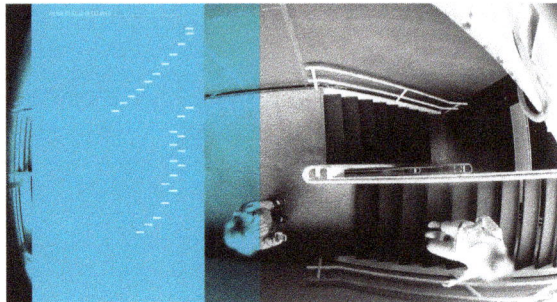

**Figure 7.** The difference between a "normal" behaviour pattern (**top**) versus a deviant behaviour pattern (**bottom**) in the Social Stairs EDL in one use session.

With over 60,000 use sessions in total the design research team found a wide range of different emergent behaviour patterns, ranging from jumping, dancing, skipping steps, trying to play a song, trying to mimic earlier patterns, etc. (Figure 8). By constructing a set of rules to determine deviant behaviour patterns from normal behaviour patterns the design research team analysed a total of 60,000 use sessions (individual behaviour patterns). Using three of the different Experiential Probes they compared the deviant behaviour patterns, both on measured data such as measured steps and time as well as clustering on the visual characteristics of the data patterns. In this analysis they found a significant difference ($p < 0.05$) between the steps and time of the use sessions of these three probes, showing a difference in the behaviour between the three different designs and their resulting emerging behaviour. Next, the team asked six fellow researchers to cluster individual use sessions on their similarities in appearance and related these clusters to the different probes through multidimensional scaling [25], through which it became clear that the deviant stair use patterns of the EP themselves differed.

**Figure 8.** The initial analysis of the data of the Social Stairs EDL at the Eindhoven University of Technology. Each sheet represents one day of data. This figure shows 13 days out of the total 6 months of data.

The design research found in this way of visualising the data, in combination with the recorded video, is a suitable way to find deviant behaviour patterns in the EDL. The data is therefore a good indicator for the design research team to see what happens in the EDL. The analysis, now done mostly by hand, has to be and can be automated to deal with these and larger quantities of data. Machine learning can here help in creating pattern recognition of the different types of deviant behaviour, as new types of behaviour constantly emerge. The interviews and user participation are however necessary to get an understanding why people are behaving the way they do.

Although we got an impression of the user behaviour in the EDL, the study also had drawbacks, mainly related to measuring behaviour change over time. Firstly, we didn't perform a baseline measurement. This is due to the fact that the Social Stairs was set up as a design exploration at first, not as a research study. By exploring and building upon design ideas the Social Stairs EP was created and set out on the go, without first measuring a baseline for a few weeks. The designers didn't know beforehand what the design would be about or that the Social Stairs would be the outcome. This highly explorative nature of the project led to the omission of a baseline measurement. The longitudinal aspect of the EDL method—the fact that Social Stairs was run for over several months—does help with creating an understanding of changes in time.

Secondly, we only measured the overall behaviour on the staircase to research if we could see different behaviour in people in relation to different behaviour/feedback on the stairs. But we did not have a set-up that enabled us to measure behaviour change on an individual level, let alone measure any long-lasting behaviour change. Our current research is moving in this direction. Our goal is to design and research systems, products and services in a way that would enable personal and social transformation. Our "design and theory for transformative qualities" aims to design and research qualities and beauty in materials and interaction, which can elevate (transform) personal/social ethics and related behaviour from an embodied interaction perspective. And with this transformation we mean long-term transformation. There are a multitude of studies that indicate, for example, the difficulty of long-lasting health-related behaviour change, such as stopping smoking, maintaining a healthy weight, or living a healthy life after a cardiac disease. As the complications of an unhealthy lifestyle are often deferred, especially with diseases such as diabetes, obesity and congestive heart failure, the motivation to adhere to therapy or a healthy lifestyle is often weak [34].

Consequently, we are looking for designs that can enable users to tap into their intrinsic motivation and connect to their own values. Now that IoT is spreading and weaving into our everyday lives, and our environment and objects are becoming interactive and smart, designs such as Social Stairs will become more regular and affordable. Our challenge is not merely to entice transformation through design, which is already huge, but also how to measure this transformation. Consequently, we are further refining the EDL method and the Data-Enabled Design approach [8] and hope to progress towards longitudinal, fully societally-embedded EDLs, in which technology enables personal and social transformation in a participatory data-enabled design process.

### 3.2. [Y]our Perspective

The second project we describe is [Y]our Perspective, which was developed within the larger "Light through Culture" educational project [35,36] lasting from 2011 to 2014. The project was promoted by the University of Siena, Eindhoven University of Technology and Interactive Institute Swedish ICT, Umeå. The aim of this educational project is to involve young designers (Bachelor's, Master's and Ph.D. students) in the development of innovative solutions for facing major social challenges of life today. The latest edition of the project was inspired by the political, economic and value-related crisis that is now affecting all of Europe. It aimed to explore how citizens can cope with the crisis and be engaged in envisioning the future of their cities. The project was developed in Siena (Italy), a city that has historically been accustomed to attributing to local communities (e.g., "contradas"; public associations) the role of promoting initiatives for safeguarding the organisation and the cultural and economical development of the territory. Unfortunately, the most important institutions in the city, which have historically supported the local community, are now experiencing a crisis that directly affects the forms of institutional arrangements, both public and private. This situation requires constant dialogue with citizens and a return to the participatory approach that characterised the life of the city in the past.

But how can citizens be actively involved in these new forms of dialogue? How is it possible to generate the right motivation, convincing citizens of the need for and importance of their participation in society? How can young people be encouraged to take part in politics, a world they now consider and perceive as some sort of parallel universe, far removed from the problems of everyday life? These are the questions the students of various disciplines (communication, design, engineering, political science) and different cultural backgrounds (Sweden, The Netherlands, Italy, Brazil, India) attempted to answer in a two-week course.

One of the design teams developed [Y]our Perspective, the design concept of an app allowing citizens and administrators to launch consultations regarding issues of public interest via smartphones. Polls are spatialised, that is, made accessible in the parts of the city involved in the consultation. For example, citizens who pass by the area where the stadium is to be built may launch a survey of people's opinions on the advisability of building a stadium in the historical centre of the city (Figure 9 left). When another person walks by the area where the stadium is to be built, the app sends the poll, in push mode (Figure 9 right). The person receiving the notification may respond to the question posed while on the site, viewing the context.

**Figure 9.** The app [Y]our Perspective. A person launches a poll about the advisability of building the new stadium in the historical centre of the city (**left**); Another person receives the poll when walking in the area (**right**).

Responses will be made available to the community, and participants will be able to state the reasons for their decisions. Responses to the poll are expressed in terms of value propositions, that is a statement of value that the person acknowledges. With the proposition, the person highlights the value or benefit that s/he would like to see and experience. Examples of value propositions are "ecology", "carbon-free", "hospitality" etc.

With reference to Marr's [7] 5 Vs of Big Data, [Y]our perspective tries to elicit mainly Variety and Value. The variety of data consists of the collection of geo-localised data, temporal data and qualitative data (topics of the pools). Value data consist of value proposition statements that highlight the personal interpretation and sense-making of people responding or proposing a poll. It is worth mentioning that the design of [Y]our perspective is conceived as a network of people gathering, but also proposing and making sense of information during the data collection. The mechanism is based on data sharing where the collection can originate from the municipality as well as from citizens. This opens up the possibility for a new dialogue between citizens and institutions.

The design of the app was driven by an investigation of different theories and methods of deliberative democracy carried out with experts' assistance.

Bobbio [37] identifies three levels of participation in democratic life. In the first level, participants merely listen to the technicians and administrators with the aim of formulating and assessing their contributions. With this approach, the administrators have the decision-making power in their hands. In the second level of participation, citizens become directly involved in the choice of plans, and are asked to debate a topic, contributing their own vision and offering the technicians an opportunity to work with the observations they have thus collected. In this case the influence of the recipients depends on the facilitator's ability to bring out the participants' demands and the ability to listen to the technicians and those promoting the process. At this level of interaction, it is still the technicians who hold the power to make decisions, and to decide whether to apply the participants' suggestions. The third level is made up of citizens who play an active role in the process, working actively with the administration, possibly with the assistance of an external facilitator, to seek a solution to a problem considered to be everyone's problem. In this case, the community plays an important and influential role, and is considered the key player in social change.

The third level includes decision-making, precisely because of the consideration and value placed on citizens' proposals and active participation. In this model, citizens become involved right away, from the earliest phases of formation of a shared political conscience, including searching for and identifying the subsequent phases of reflection on the themes of the debate. Therefore, this model potentially offers a way to encourage citizens to cooperate in the search for solutions to social and political dilemmas, instigating active involvement which does not stop at the choice of people to represent them in political elections.

On the basis of this theoretical framework as well as interviews conducted with citizens on issues of common interest, and of the study of examples of good government taken from the history of the city of Siena, students developed a video concept that was used as the basis for participatory design sessions with Sienese citizens. The video is included in a larger video documentary available at [38].

The design of the poll mechanism of the app was inspired by the "Conflict Spectrum", a methodology used in deliberative democracy to solve conflicts at an early stage [39]. The method allows people to understand the motivations of other people taking a certain position as well as the number of people sharing certain opinions. It requires that participants publicly express their point of view which is used only when people feel comfortable enough to take a stand openly. The method works like a bodystorming. Participants have to align themselves along a virtual line connecting two corners in the room. People who are firmly convinced of a certain position move to one corner. People who are convinced of the opposite move to the opposite corner. The participants are informed that infinite shades may exist between the two positions represented in the room by intermediate positions between the two corners. Each participant takes the position in the room corresponding to his/her view, and when people are placed along the spectrum, the facilitator asks them to briefly explain why they chose to get in that position. During the explanation, people are allowed to change their position in the spectrum toward a more convincing stand. If anyone chooses a new location, s/he is invited to explain why. The expected result is a better understanding of the personal opinion and that of the others, the acquisition of a more collaborative and open stand towards the opinion of others and hopefully the conflict resolution.

Inspired by this method, the graphical interface of the app was designed as a virtual conflict spectrum with opposite opinions placed at the extremes of a line (Figure 10). The person responding to a poll can position her opinion along the line, expressing also the values or expected benefits from voting that position. He/she can also read other responses to the poll. These data are stored in the system and the results of the pool are dynamically visualised on a public large screen display in the city hall.

**Figure 10.** The graphical interface of [Y]our Perspective inspired by the Conflict Spectrum.

From the perspective of transformation, both empowerment and recognition of the other are strongly present in the design of this app. The spectrum empowers by offering virtual space for expressing the own voice, giving to the citizens a concrete, spatial frame of reference in which to locate their beliefs and points of view. It also creates an implicit sense of connection to the others: even those who stand at the opposite end of the spectrum stand in a continuum with the others. Also, cooperation with one's opponent in a joint activity dedicated to elucidating the views through value propositions in a positive and constructive way, creates a temporary ritual of common purpose.

[Y]our Perspective platform was designed through a number of participatory design activities that involved designers, students, experts in the field of political science, citizens and representatives of local public administration. The process did not however include any behavioural data sensing, mining or analysis with the actual prototype which served the purpose of consolidating the design concept rather than actually collecting data on the field through the platform.

However, participatory design sessions involved a large number of stakeholders (up to 150 as explained below) and highlighted a number of issues that clearly show the importance of a participatory approach to data-enabled design. First of all the interviewees stated that they would use the system only in the event that the municipality ensures that the measures discussed and approved by the citizens were actually placed on the political agenda. In this way, users would be strongly encouraged to contribute by using the app and feel they would want to continue to do so in the long term. In other words, gathering data without the chance to see it implemented was not seen to be worthwhile.

Yet it appears that the use of the app would be appreciated and spread if it turns out to be a social platform for communication and sharing, rather than a simple poll used by the municipality to monitor the citizenship's opinions. People do not feel at ease generating data without taking part in the interpretation and implementation process. The fact that both parties (citizens and municipality) can propose a topic is a valuable point, because in this way the system would not be seen as a means for the public administration to achieve political consensus, but as a real attempt at communication and listening.

Another aspect expressed by almost all participants of the participatory design sessions is the importance of continuing the online discussions in public meetings with other citizens and local government representatives. A proposal should not halt at the level of exchanging views, but should result in subsequent collective deliberation, subject to the monopoly of the government on specific issues. This suggestion was elaborated a few months later in a follow-up project consisting in a large public consultation organised in Siena in the form of an Open Space Technology (OST) meeting involving around 150 participants (Figure 11). Some of the themes discussed during the OST have been recognised as priorities in the political agenda of the municipality of Siena.

The OST served as a means to simulate the sense-making process that the app could stimulate on the basis of the data gathered on different pools.

**Figure 11.** The public consultation organised in Siena.

## 4. Conclusions

In this paper, through two projects we elucidated our approach to data-enabled design as a social proactive participatory design approach. In line with the paper from Bogers et al. [8] we see three important building blocks for a data-enabled design approach: user involvement in participatory co-design processes, methodologies for technology-mediated data collection from the field, and the use of interactive prototyping as a vehicle for experience and value probing and a means to obtain data starting early on in the design process.

Our projects and the underlying EDL method were built on the notion that the strength of these building blocks can be found in their synergy. Using processes of (1) Envisioning; (2) Designing interventions; (3) Acquiring data; and (4) Analysing and modelling, we have tools to start tackling societal challenges together with other stakeholders. We believe that the design of smart systems would benefit from a data-enabled design approach considering big data about personal and social behaviour patterns from a user-centred perspective. This perspective sees people not only as data gatherers but as active participants in the process of interpreting and making sense of data. That is why we frame our approach as data-enabled design, which includes a participatory/co-design approach with a high level of prototyping (from lo-fi to hi-fi) throughout the process, thus developing wise systems that acknowledge and include our sociocultural context. In the data-enabled design approach we see fluent connections between the physical and the digital, the technical and the socio-cultural, the designers/developers and the citizens/customers, and between synthesis and analysis.

The projects that we show in this paper are merely a start to explore this new and exciting road towards data-enabled design, where technology is used to bring people and communities at the core of ecosystems of interconnected products and services, thus finding a more successful way to cope with the current complexity and our societal challenges. By developing new methods and approaches such as the EDL, which aims to transform data-driven design in a participatory and co-creative data-enabled design approach, we hope to offer others refreshing ways to design "smart" systems. Given the new prosumers-drive in the knowledge economy, we like to make it as concrete, open and accessible as possible and invite others to experiment with it.

**Acknowledgments:** We would like to thank the students of the Eindhoven University of Technology and the University of Siena for their enthusiastic and dedicated participation in the projects described in this paper. Many people actively contributed to the two projects and we would like to sincerely thank them. For Social Stairs: Michel Peeters, Sander Bogers, Aarnout Brombacher, Mathias Funk, Jean-Bernard Martens, Nick Hermans, Bart Wolfs, Rhys Duindam, Max Sakovich, Nadine van Amersfoort, Stijn Stumpel as well as the students and colleagues of the Eindhoven University of Technology who used and tested the Social Stairs. We thank Koninklijke KPN for enabling us to build a version for their headquarters. For [Y]our Perspective: Pierangelo Isemia, Iolanda Romano, Nigel Papworth, Ambra Trotto, Evert Wolters, Iolanda Iacono, Rosa De Piano, Michele Tittarelli, Arvid Jense, Trieuvy Luu as well as the students from the Eindhoven University of Technology, University of Siena and Interactive Institute Swedish ICT Umeå, who participated in this Light through Culture project. Finally, we like to thank Comune di Siena, the Archivio di Stato di Siena, the Contrada della Lupa and Necker van Naem for their support during this project.

**Author Contributions:** The authors equally contributed to the paper defining the concept of participatory data-enabled design. Caroline Hummels and Patrizia Marti conceived the educational project "Light through Culture" and coordinated the educational programme for four years (2011–2014). They supervised the students' project [Y]our perspective, organised the participatory sessions in the city of Siena and contributed to the data analysis and interpretation. Carl Megens created the Experiential Design Landscapes Master's course for three years, from which the Social Stairs project emerged. This project was then further developed for research purposes.

**Conflicts of Interest:** The authors declare no conflict of interest.

## Abbreviations

The following abbreviations are used in this manuscript:

| | |
|---|---|
| EDL | Experiential Design Landscape |
| EP | Experiential Probe |
| OST | Open Space Technology |
| PS | Participatory Sensing |

## References

1. Hienz, J. The Future of Data-Driven Innovation. U.S. Chamber of Commerce Foundation, 2014. Available online: https://www.uschamberfoundation.org/future-data-driven-innovation (accessed on 16 September 2016).
2. Gardien, P.; Deckers, E.; Christiaansen, C. Innovating Innovation—Deliver meaningful experiences in ecosystems. *DMI Acad. Des. Manag. Conf. Lond.* **2014**, *9*, 36–46.
3. Brand, R.; Rocchi, S. Rethinking Value in a Changing Landscape: A Model for Strategic Reflection and business Transformation. *Eindh.: Philips Des.* **2011**. Available online: http://www.design.philips.com/philips/shared/assets/design_assets/pdf/nvbD/april2011/paradigms.pdf (accessed on 16 September 2016).
4. Swan, M. The Quantified Self. *Big Data* **2013**, *1*, BD85–BD99. [CrossRef] [PubMed]
5. Tomico, O.; van Zijverden, M.; Fejér, T.; Chen, Y.; Lubbers, E.; Heuvelings, M.; Schepperheyn, V. Crafting wearables: Interaction design meets fashion design. In *CHI'13 Extended Abstracts on Human Factors in Computing Systems*; ACM: New York, NY, USA, April 2013; pp. 2875–2876.
6. Weber, R.H.; Weber, R. *Internet of Things*; Springer: New York, NY, USA, 2010; Volume 12.
7. Marr, B. *Big Data: Using SMART Big Data, Analytics and Metrics to Make Better Decisions and Improve Performance*; John Wiley & Sons: Hoboken, NJ, USA, 2015.
8. Bogers, S.; Frens, J.; Kollenburg, J.; Deckers, E.; Hummels, C. Connected Baby Bottle: A Design Case Study towards A Framework for Data-Enabled Design. In Proceedings of the Designing Interactive Systems Conference, Brisbane, Australia, 4–8 June 2016.
9. Marti, P.; Bannon, L. Exploring User-Centred Design in Practice: Some Caveats. *Knowledge Technol. Policy* **2009**, *22*, 7–15. [CrossRef]
10. Acki, P.; Honicky, R.; Hooker, B.; Mainwaring, A.; Myers, C.; Paulos, E.; Subramanian, S.; Woodruff, A. A Vehicle for Research: Using Street Sweepers to Explore the Landscape of Environmental Community Action. Available online: http://www.urban-atmospheres.net/CitizenScience/ (accessed on 9 August 2016).
11. Paulos, E.; Honicky, R.; Hooker, B. Citizen Science: Enabling Participatory Urbanism. In *Handbook of Research on Urban Informatics: The Practice and Promise of the Real-Time City*; Foth, M., Ed.; Information Science Reference, IGI Global: Hershey, PA, USA, 2009.
12. Goldman, J.; Shilton, K.; Burke, J.; Estrin, D.; Hansen, M.; Ramanathan, N.; Reddy, S.; Samanta, V.; Srivastava, M; West, R. *Participatory Sensing: A Citizen-powered Approach to Illuminating the Patterns that Shape Our World (White Paper)*; Woodrow Wilson International Center for Scholars: Washington, DC, USA, 2009.
13. Participatory Sensing. Available online: http://www.mobilizingcs.org/about/participatory-sensing (accessed on 16 September 2016).
14. Blum, J.; Flintham, M.; Jacobs, R.; Shipp, V.; Kefalidou, G.; Brown, M.; McAuley, D. The Timestreams platform: Artist mediated participatory sensing for environmental discourse. In Proceedings of the 2013 ACM International Joint Conference on Pervasive and Ubiquitous Computing (UbiComp '13), Zurich, Switzerland, 8–12 September 2013; pp. 285–294.
15. Sanders, E.B.N.; Stappers, P.J. Co-creation and the new landscapes of design. *CoDesign* **2008**, *4*, 5–18. [CrossRef]
16. Batch, B.; Tyson, C.; Bagwell, J.; Corsino, L.; Intille, S.; Lin, P.; Lazenka, T.; Bennett, G.; Bosworth, H.; Voils, C.; et al. Weight loss intervention for young adults using mobile technology: Design and rationale of a randomized controlled trial. Cell Phone Intervention for You (CITY). *Contemp. Clin. Trials* **2014**, *37*, 333–341. [CrossRef] [PubMed]
17. Hine, C. Virtual ethnography: Modes, varieties, affordances. In *SAGE Handbook of Online Research Methods*; SAGE: New York, NY, USA, 2008; pp. 257–270.
18. Bolt, N.; Tulathimutte, T. *Remote Research*; Rosenfeld Media, LLC: New York, NY, USA, 2010; p. 266.
19. Gardien, P.; Djajadiningrat, J.P.; Hummels, C.C.M.; Brombacher, A.C. Changing your hammer: The implications of paradigmatic innovation for design practice. *Int. J. Des.* **2014**, *8*, 119–139.
20. Tam, D. Facebook by the Numbers: 1.06 Billion Monthly Active Users. Available online: http://news.cnet.com/8301-1023_3-57566550-93/facebook-by-the-numbers-1.06-billion-monthly-active-users/ (accessed on 25 October 2013).

21. Watkins, T. Suddenly, Google Plus is Outpacing Twitter to Become the World's Second Largest Social Network. Available online: http://www.businessinsider.com/google-plus-is-outpacing-twitter-2013-5# ixzz2aNZf9cYb (accessed on 16 September 2016).
22. De Jaegher, H.; di Paolo, E. Participatory sense-making: An enactive approach to social cognition. *Phenomenol. Cogn. Sci.* **2007**, *6*, 485–507. [CrossRef]
23. Dewey, J. *Experience and Education*; Touchstone: New York, NY, USA, 1938.
24. Schön, D. *The Reflective Practitioner*; Basic Books: New York, NY, USA, 1983.
25. Peeters, M.; Megens, C. Experiential Design Landscapes: How To Design For Behaviour Change, Towards an Active Lifestyle. Ph.D. Thesis, Technische Universiteit Eindhoven, Eindhoven, The Netherlands, 15 April 2014.
26. Peeters, M.; Megens, C.; Ijsselsteijn, W.; Hummels, C.; Brombacher, A. Experiential Design Landscapes: Design research in the wild. In Proceedings of Nordic Design Research Conference, Copenhagen, Denmark, 9–12 June 2013.
27. Peeters, M.; Megens, C.; van den Hoven, E.; Hummels, C.; Brombacher, A. Social Stairs: Taking the Piano Staircase towards Long-Term Behavioural Change. *Persuas. Technol.* **2013**, *7822*, 174–179.
28. Owen, H. *Open Space Technology: A User's Guide*, 3rd ed.; Berrett-Koehler: Oakland, CA, USA, 2008.
29. Ross, P.; Tomico, O. The Growth Plan: An approach for considering social implications in Ambient Intelligent system design. In Proceedings of the AISB 2009 Convention, Edinburgh, Scotland, 6–9 April 2009.
30. Verbeek, P.P. Materializing morality design ethics and technological mediation. *Sci. Technol. Hum. Values* **2006**, *31*, 361–380. [CrossRef]
31. Ihde, D. *Technology and the Lifeworld: From Garden to Earth*; Indiana University Press: Bloomington, IN, USA, 1990; No. 560.
32. MacCallum, L.; Howson, N.; Gopu, N. *Designed to Move: A Physical Action Agenda*; NIKE: Kokkedal, Denmark, 2012.
33. Van der Ploeg, H.P.; Chey, T.; Korda, R.J.; Banks, E.; Bauman, A. Sitting time and all-cause mortality risk in 222 497 Australian adults. *Arch. Intern. Med.* **2012**, *172*, 494–500. [CrossRef] [PubMed]
34. Christensen, C.; Grossman, M.; Hwang, J. *The Innovator's Prescription; How Disruptive Innovation Can Transform Health Care*; The McGraw-Hill Companies: New York, NY, USA, 2009.
35. Marti, P.; Overbeeke, K. Designing Complexity in Context: Light through Culture. In Proceedings of the Ninth Conference of the Italian Chapter of ACM SIGCHI (Association for Computer Machinery—Special Interest Group on Computer-Human Interaction), Alghero, Italy, 13–16 September 2011; pp. 65–70.
36. Marti, P.; Trotto, A.; Peters, J.; Hummels, C. Instilling Cultural Values through Bodily Engagement with Human Rights. In Proceedings of the Tenth Conference of the Italian Chapter of ACM SIGCHI (Association for Computer Machinery—Special Interest Group on Computer-Human Interaction), Trento, Italy, 16–19 September 2013.
37. Bobbio, L. *A Più Voci. Amministrazioni pubbliche, Imprese, Associazioni e Cittadini nei Processi Decisionali Inclusivi*; Edizioni Scientifiche Italiane: Roma, Italy, 2004.
38. Design for Politics LtC. Available online: http://vimeo.com/98062969 (accessed on 16 September 2016).
39. Kraybill, R. *Facilitation Skills for Interpersonal Transformation, Berghof Handbook for Conflict Transformation*; Berghof Research Center for Constructive Conflict Management: London, UK, August 2004.

![future internet logo] *future internet*

|MDPI|

*Article*

# Computational Social Science, the Evolution of Policy Design and Rule Making in Smart Societies

Nicola Lettieri [1,2]

1   ISFOL, Institute for the Development of Vocational Training, Corso d'Italia 33, 00198 Rome, Italy;
    n.lettieri@isfol.it; Tel.: +39-0824-351-232
2   Department of Law, Economics, Management, Quantitative Methods, University of Sannio, Piazza Arechi III,
    82100 Benevento, Italy

Academic Editor: Dino Giuli
Received: 12 March 2016; Accepted: 27 April 2016; Published: 12 May 2016

**Abstract:** In the last 20 years, the convergence of different factors—the rise of the complexity of science, the "data deluge" and the advances in information technologies—triggered a paradigm shift in the way we understand complex social systems and their evolution. Beyond shedding new light onto social dynamics, the emerging research area of Computational Social Science (CSS) is providing a new rationale for a more scientifically-grounded and effective policy design. The paper discusses the opportunities potentially deriving from the intersection between policy design issues and CSS methods. After a general introduction to the limits of traditional policy-making and a brief review of the most promising CSS methodologies, the work deals with way in which the insights potentially offered by CSS can concretely flow in policy choices. The attention is focused, to this end, on the legal mechanisms regulating the formulation and the evaluation of public policies. Our goal is two-fold: sketch how the project of a "smart society" is connected to the evolution of social sciences and emphasize the need for change in the way in which public policies are conceived of, designed and implemented.

**Keywords:** computational social science; policy modeling; rule making; smart society; social simulation

---

## 1. Introduction: Policy Failures as Failures of Knowledge

In a talk given in 2005, while presenting the *Committee on Global Thought* (New York, NY, USA, 14 December 2005), a commission intended to build an international program for the study of globalization and its issues, Lee Bollinger, the Dean of Columbia University said:

> "The forces affecting societies around the world and creating a global community are powerful and novel. The spread of global market systems, the rise of (and resistance to) various forms of democracy, the emergence of extraordinary opportunities for increased communication and of an increasingly global culture, and the actions of governments and nongovernmental organizations are all reshaping our world and our sense of responsibility for it and, in the process, raising profound questions. These questions call for the kinds of analyses and understandings that academic institutions are uniquely capable of providing. Too many policy failures are fundamentally failures of knowledge..." [1]

As suggested by Bollinger's words, humankind is dealing with a series of novel and global challenges spanning from the depletion of natural resources to migrations; from financial, institutional and economic instability to the spreading of epidemics; from international terrorism to corruption or illegal use of data. We live in a complex world in which the interactions between technological, economic, social and political systems are becoming ever more frequent and mazy: in our hyperconnected society, any event can produce effects that spread rapidly from one system to another through cycles of nonlinear feedback that are extremely difficult to predict and to control [2].

In this scenario, the ambitious project of a "smart society" [3], the vision of a socio-technical ecosystem exploiting advanced information and communication technologies to build resilient institutions [4], improve the quality of life and solve the severe challenges of modernity, encounters serious difficulties. Recent global financial and economic crises have casted grave misgivings on the robustness and the scientific basis of policy-makers' choices emphasizing the inadequacy of our understanding of the mechanisms governing social complexity.

Often characterized by a naively mechanistic vision of reality and by the underestimation of the interaction between public choices and individual decision-making, the conventional approaches to policy modeling show huge limitations in terms of efficacy [5]. On a closer examination, the situation appears to be the result of different factors, two of which are particularly relevant for our analysis insofar as expression of the theoretical and scientific deficiencies of conventional policy-making.

The first one [6] is represented by the difficulty to grasp the logic underlying human decision-making. Individuals do not necessarily behave in the linear ways predicted by rational choice theory [7], which still is the main source of inspiration for policy-makers. Even when dealing with the same information or the same payoffs, human beings show behaviors that are often hard to anticipate.

The second reason has to do with the fact that, differently from what is predicted by economic theory, the preferences of individuals are continuously altered by the interaction with others. Even knowing in advance the reaction of an individual to a policy change, it is impossible to foresee the evolution of his behavior as the response depends also on the reaction of the other members of the same group and on learning mechanisms that are anything but linear. Strongly amplified by the interactions taking place in real time on large-scale communication networks, the adaptive component of social behaviors generates tricky implications that are difficult to deal with [5].

Luckily, in the last two decades, the scientific understanding of the social and economic world has started to make relevant progress with respect to the above-mentioned issues. Social scientists have begun to investigate social systems with new theoretical and technological tools allowing innovative approaches to social research. A fundamental role has been played, in this vein, by two different scientific advancements. If, on the one hand, Behavioral economics and Economic psychology [8,9] started to offer new insights into how individuals and groups decide, interact and react to changes, on the other hand [6], the conceptual framework offered by complexity theory, the data deluge [10] and the computational evolution of science have paved the way to an unprecedented leap in the comprehension of social phenomena.

This paper discusses how the advancements of social sciences can contribute to the challenge of a smarter society providing policy analytics with a new theoretical and instrumental framework. The attention is focused, in particular, on the scientific and methodological acquisitions gained in the emerging area of Computational Social Sciences (CSS), a data- and computation-driven research field that is going to play (this our main claim) a crucial role in promoting a more scientifically-grounded approach to policy design and a systemic vision of public choices.

The article is structured as follows. The first section briefly introduces, in general terms, the computational evolution of science. The second and the third sections analyze the computational social science paradigm and its main research methodologies. The fourth section discusses the intersections between computational social science and policy modeling, while the fifth is focused on the role that can be played, in this scenario, by social simulation models. The last section concludes the paper, sketching some final considerations.

## 2. The Data (and Computation)-Driven Future of Science

The advancement of science has always been strongly influenced by the features of the research tools used by the scientist. For a long time supported only by the direct observation of reality, the scientific endeavor has been over the centuries mediated by increasingly complex artifacts capable of offering new and more realistic representations of the world. As a matter of fact, research is an "instrument-enabled" [11] activity, the result of an iterative process in which technological development is at the same time an enabling factor and an outcome of scientific progress. On this process depended the birth of entire research areas: the nano-sciences, to give just a recent example, would not have come into being without the invention of the electronic microscope.

Much the same applies to information technology. Digitization, the key prerequisite allowing a computer to process information, is a decisive element for scientific research in the broadest sense understood, and we are going through a period of dramatic growth of digital information. In the era of Big Data, a process fueled by the web and social media, as well as by the spread of smartphones, surveillance devices and increasingly large networks of sensors, the collection of a huge amount of data is allowed, from which you can extract knowledge with effects that take place both on the scientific and the application level [12,13].

This circumstance is highlighted in a recent report by the Royal Society of London (London, England) [14], the oldest British scientific association, which analyzes the increasingly close link between science and the massive analysis of digital data. By leveraging the growing computational power today available and algorithms capable of more and more reliable inferences, computational sciences allow one to extract knowledge existing only implicitly within large datasets. In this scenario, data are used not only to validate theories and hypotheses, but also to identify regularities and correlations that are useful both to better understand reality and to make predictions about the future.

We are facing what has already been defined [15] as a new research paradigm that is adding to the existing ones. Since the 17th century, scientists have identified in theory and in experimentation the two basic scientific paradigms allowing human beings to understand nature. In last few decades, computer simulations have become the third paradigm of science, the standard way to explore phenomena that are inaccessible to theory and experiments, from climate change to the evolution of galaxies.

Today, "data-led" science, the science that exploits the analysis of huge amounts of data to produce knowledge [16], represents the fourth scientific paradigm, an emerging approach that can already count on innovative practices that span from "data mining" (the discovery of pattern and regularities within large sets of digital data) to "information integration" (the merging of information coming from heterogeneous sources with differing conceptual, contextual and typographical representations), from speech to image recognition.

### 3. From Nature to Social Facts: Computational Social Science

The above described development is not going to remain within the borders of the physical and natural sciences that are, for understandable reasons, more familiar with data analysis and computation. Only six years ago, *Science* published a position paper [17] that identified in the digital information flood and in the rise of computer capacities the starting point for a scientific and methodological renewal expected to heavily impact social sciences. In the perspective of "Computational Social Sciences" (CSS), a research area that has experienced an almost exponential growth over the last 10 years (see Figure 1), the analysis of data generated by our interactions with digital technologies, the use of quantitative and computational methods of investigation together with new forms of experimentation push human sciences to the cumulativeness and the rigor that have historically characterized the study of the physical and biological world.

The new ICT-enabled study of society is grounded in a strongly interdisciplinary approach to the investigation of social systems: scientists belonging to different research areas ranging from cognitive and behavioral science to physics and economics strictly cooperate to come up with innovative models of social phenomena. On the one hand, they take advantage of massive ICT data representing traces of almost all kinds of activities of individuals. On the other, they exploit the growth of computational power and the modelling instruments made available by ICT to produce predictive and explanatory models of social systems [18,19].

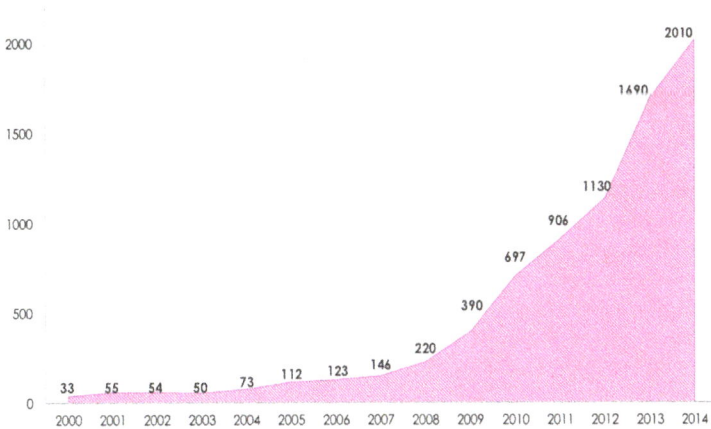

**Figure 1.** Number of papers published in the period 2000–2014 and containing an explicit reference to "computational social science" in the title (source: Google Scholar).

Although relatively new, computational social science is already [11,19] a well-established research area organized in an international and active community that by now covers many topics. The computational modeling of social phenomena and the analyses of huge datasets obtained from mobile social networks, phone calls or online commercial interactions have already started to shed new light onto extremely various subjects like political opinions [20,21], the evolution of individual tastes [22], the relationships between the structure of society and the intensity of relationships [23], the spread of pandemic diseases [24] or the speed of communication [25].

From a theoretical and epistemological point of view [26,27], computational social science, the "integrated, interdisciplinary pursuit of social inquiry with emphasis on information processing and through the medium of advanced computation" [11], is grounded on a scientific perspective in which different research traditions flow into one (see Figure 2). The CSS scientific background is extremely varied and encompasses contributions from complexity science, general system theory, cybernetics, information theory, computer science, mathematics, economics, sociology, political science and many more. The theoretical foundations of the area can be therefore traced back to the seminal works of heterogeneous scientists, like Ross Ashby, Claude Shannon, Norbert Wiener, Herbert Simon, Ludwig von Bertalanffy, Harold Guetzkow and Thomas Schelling [28].

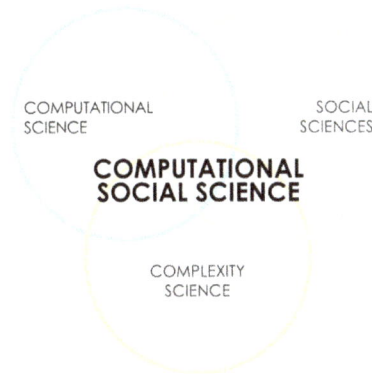

**Figure 2.** Computational social science: scientific background (source: adapted from ref.[11]).

From a methodological standpoint, computational social science can rely on an already mature framework that encompasses different research methods with several specialized sub-branches. According to Cioffi-Revilla [19], if we drop for now visualization [29], visual analytics [30] and sonification [31], which will likely become autonomous research practices in the near future, we can identify five main CSS methods that contribute in different ways to the computational investigation of social matter: information extraction; complexity models; social network analysis; geographic information systems; social simulation models. Several combinations among the main five methods (like network analysis, GIS and social simulation; see, just for instance, [31]) are already common. Others, since the field is still young, have yet to be explored. As we briefly highlight here below, each of the mentioned methods offers specific insights allowing to see beyond the scope of traditional social science research practices or even beyond the results offered by earlier statistical approaches.

*Making sense of data: information extraction and data mining.* The extraction of semantically-meaningful information from unstructured documents and data sources is probably the keystone of computation-driven science. The identification of unsuspected pattern, regularities and correlations within huge amounts of data allows not only to produce new knowledge by means of inferences and predictions, but is also a powerful incentive for theory making. Born as the application of parsing and marking (semantic encoding) techniques to textual database, information extraction has evolved into the increasingly sophisticated computational analysis of audio, images and videos. A significant step forward in this direction has been made with artificial intelligence methods that are already producing promising results. Automated textual analysis represents a valuable strategy in all of the areas of the social sciences in which the understanding of the text has a crucial scientific role. One of the primary uses of automated information extraction is the production of series of *events data*, a stream of historical, social or economic information that can be analyzed through methodologies spanning from more traditional statistical [32] time series to event lifecycle analysis [32]. Another relevant application of automated text mining [33] is the generation of network data structures analyzable with social network analysis techniques (see, for instance, [34]). In applied research fields, automated information extraction can be used not only for the detection of anomalies and early warning, but also to analyze trends of social and economic dynamics. This feature could turn out to be useful to monitor the impact of policy intervention, especially when dealing with real-time data streams (e.g., stock exchange data).

*Grasping (and predicting) the behavior of complex systems.* In the last twenty years, complexity theory has provided social science with powerful methods for the analysis of non-equilibrium dynamics that are quite often found in many social phenomena [35–37]. Based on mathematical concepts and principles, complexity-theoretic models illuminate the rules underlying the behavior of social systems that, like all complex systems, are characterized by non-linear and hardly predictable dynamics. Social scientists are provided with the opportunity to draw important inferences about the potential evolution of social dynamics that are neither available nor reliable on the basis of data or plain observation [37]. These inferences include (but are not limited to) the risk of extreme events, the fragility of unstable conditions or the early-warning indicators of incumbent drastic changes. Patterns observed in terrorist attacks, in the distribution of poverty and wealth of developing countries, political instability or, again, the dynamics of the encounter between supply and demand in the labor market are typical examples of nonlinear and non-equilibrium processes. Complexity models allow one to find the regularity of these dynamics, to understand and often predict their evolution [38]. That is why many relevant social dynamics have been investigated by means of the "complexity-inspired" approach from market fluctuations [39] to extremist religious opinions [40] through a process that is gradually extending to many other social issues.

*From (social) structure to function: social network analysis.* The networks, structures made of entities (nodes) and relations between entities (links) each of which is defined by specific properties, are recurring in a large amount of social phenomena. Communities, groups, political parties, criminal organizations and international alliances are common examples of networks of interest to social scientists. Social Network Analysis (SNA) is a research methodology rooted in social psychology and in the mathematical theory of graphs that exploits formal and computational methods to investigate the structural and functional properties of social networks. Thanks to a large family of metrics and methods, SNA allows to extract insightful information and to make meaningful inferences about the features of a social network simply starting from the analysis of its structural pattern of nodes and relations [41]. The investigation develops in two phases: the construction of a graphical representation (graph) of data related to the target social phenomenon and the quantitative analysis of the graph based on standard methods and metrics. Through this process, network properties, such as resilience, vulnerability and functionality, can be inferred in a way that is not even conceivable with plain observation or more traditional social science methods (surveys, questionnaires or even statistical analysis). As emerges by a growing literature, SNA has already had large applications across the social sciences, providing a deeper understanding of many different social phenomena: language diffusion; opinion dynamics; belief systems (e.g., political ideologies); treaty systems and their historical evolution; international and transnational organizations like terrorist networks [42,43]. Moreover, SNA can be applied to the design of robust and sustainable networks in areas of primary importance for policy-makers, like infrastructures and public transport, security and health.

*Spatial analysis of social phenomena*. Geographic Information Systems (GIS) are systems designed to store, integrate, analyze, share and display information placed in spatially-/geographically-referenced contexts. GIS were first introduced in social research by social geographers to visualize and analyze spatially-referenced data about the social world. Nowadays, *SocialGIS* allows one to investigate the spatial dimension of social phenomena being applied in various fields of social sciences, from criminology [31] to the regional economy [44], often in conjunction with other quantitative techniques. In general terms, the methodology is characterized by the capacity to offer a synoptic view of various categories of social data thanks to the superposition of several levels of information (as occurs, for example, in Google Earth). Thanks to these features, *SocialGIS* can be a valuable ally for the policy-maker, as it enables one to monitor in innovative ways the scenarios object of intervention and the impact of public choices.

*Reproduce to explain (and predict): social simulation models*. Scientific explanation is strictly connected with experiments. They not only push to explicitly formulate hypotheses about the factors and the processes that can causally explain a given phenomenon, but also allow one to validate the assumption by means of the comparison between the experiments' results and the empirical predictions made by the researcher. Social simulations are computational models that provide social scientists with the opportunity to explain in experimental terms and to predict the evolution of complex social phenomena reproducing the real-world processes that generate the social reality. The two main simulation techniques used today in social research are *Systems dynamics* [45] and *Agent-Based Models* (ABM) [46,47]. *System dynamics models* are based on the idea that the evolution of a social system can be represented as the result of complex cycles of action and feedback which can be described in mathematical terms. On this assumption are grounded simulations in which the phenomenon under investigation is represented as a set consisting of variables (*stocks*) and rates of change (*flows*) associated with them. Today, these models are the basis for many applications in the industrial sector, managerial and social sciences. *Agent-based simulation models* are led by the theoretical assumption that the macro-level social phenomena (e.g., the emergence of social norms or the spread of racial segregation) are the result deriving from the interactions occurring, at the micro-level, between individuals and between individuals and the environment. ABMs typically include a set of heterogeneous artificial agents simulating real-world actors and their behaviors, a set of rules of interaction and an environment in which both dynamic, organizational and spatial characteristics are defined. As highlighted in the next section, social simulation provides policy-makers with the ability to run a *what-if analysis* allowing to observe the effects potentially deriving from different choices by means of well-developed models of a given "target system" (see, for example, the effects of social stressors on patterns of political instability [48]).

## 4. Computational Social Science and Policy Design: Making Smarter Public Choices

The implications of the scenario so far described go beyond the scientific dimension. By increasing our ability to understand and predict social dynamics, computational social science is preparing the ground to innovative and more rooted in science approaches to policy design. Thanks to the capacity of bringing together in innovative ways theory and computation, data and laboratory experiments, computational social science offers policy makers and legislators a clearer idea of complex socio-economic processes and an opportunity to increase the effectiveness of public choices. CSS methodologies are particularly suited to the study of non-linear phenomena that are, at the same time, difficult to understand with conventional mathematical and statistical tools and also often poorly understood by traditional policy-making procedures. That is why, in recent years, the scientific community has paid a growing attention to the intersections between computational social sciences and policy modeling [2,12,19].

As evidence of this, we can cite some research projects that explore the potential impact of computational methodologies in policy-making and try to promote what the EU defines as *"scientific evidence-based policy making"* [49], drawing attention to the need for policy-makers to ground their decisions in science. *FuturICT* [50] is a research project presented in response to the FP7 Flagship Call ICT aiming to bring together complexity science, social sciences and information technologies to support the understanding and the management of complex global issues spanning from epidemics to crime, ecological disasters and financial crises. Largely inspired by CSS research, the core of the project is represented by the *Living Earth Platform*, an advanced simulation, visualization and participation platform designed to support decision-making of policy-makers, business people and citizens. A similar issue, the understanding and the prediction of systemic risk and global financial instabilities is the main topic of *FOC - Forecasting Financial Crises* [51], a project financed by the Future Emerging Technologies OPEN Scheme aiming, on the one hand, to offer a theoretical framework to measure risks in global financial market and, on the other hand, to deliver an ICT collaborative platform for monitoring systemic fragility and the propagation of financial distress across markets around the world.

Dealing with the economy is also *Eurace* [52], FP6 European research that involved researchers from various backgrounds to build an agent-based simulation model of the European economy. The research tackles the complexity generated by the interplay between different macroeconomic policies (fiscal and monetary strategies, R&D incentives, *etc.*) pursuing objectives that unfold on both the scientific and the societal level. From the scientific point of view, the main goal was the development of multi-agent models reproducing the emergence of global features of the economic system from the complex pattern of interactions among heterogeneous individuals. From the social point of view, the goal was to support EU policy design capabilities by means of simulation enabling one to perform what-if analysis, optimizing the impact of regulatory decisions that will be quantitatively based on the European economy scenario.

*FUPOL* (FUture POLicy modelling) [53], finally, is an FP7 project that aimed to provide a new approach to politics building on major innovations like multichannel social computing, crowd sourcing and social simulation. The project developed a governance model to support the policy design and implementation lifecycle by a system combining two CSS techniques: information extraction and social simulations. The first was used to collect, analyze and interpret the opinions expressed online by citizens, in order to give governments a better understanding of their needs. The latter aimed to simulate the effects of policies in order to assist governments throughout the process of their development.

Beyond the albeit significant scientific results, marked by a consistent number of relevant publications [2,19,54,55], the above-mentioned projects share the credit of enhancing the debate over the intersections between policy modeling and science promoting, at the same time, the creation of an interdisciplinary research community. As the result of their activities somehow shows, CSS research is gradually driving a deep innovation in policy design. The methodologies and the tools developed by computational social scientists seem indeed to be helpful in different fundamental steps of the policy cycle from the identification of the solutions that are more appropriate to implement the policy to decision-making and *evaluation* (see Figure 3), opening up new prospects and new scenarios.

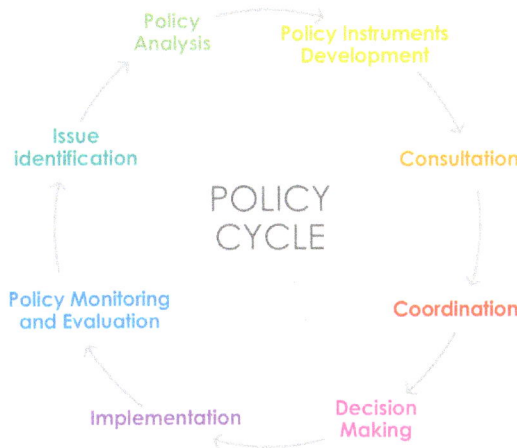

Policy
Analysis                    Policy Instruments
                            Development

Issue
identification                              Consultation

                    POLICY
                    CYCLE

Policy Monitoring                           Coordination
and Evaluation

        Implementation              Decision
                                    Making

**Figure 3.** The policy cycle (source: adapted from ref.[56]).

A reasonable question at this point is: how can the insights potentially deriving from CSS methodologies concretely be implemented into the policy design process? A crucial role is expected to be played, in this vein, by the rule-making procedures and, in more detail, by the legal mechanisms regulating the formulation and evaluation of public policies. The issue is strictly connected with the goal of a higher 'quality of regulation' [57], an increasingly important concern on the agenda of national and international institutions.

In the last twenty years, the Organisation for Economic Co-operation and Development (OECD) has produced several documents on the topic adopting, in 2012, a document (*"Recommendation on regulatory policy and governance"*) [58] that contains detailed international guidelines on the quality of regulation. The EU, in its turn, has been moving in the same direction actively promoting the creation of tools to ensure the the the quality of norms discussing, since the early 1990s, new ideas and practices to assess the impact of both policies and rules implementing them. More recently, the subject has been widely analyzed within the context of the White Paper on Governance [59] in which the European Commission expressly emphasizes the need to improve the quality, the simplicity and the effectiveness of regulatory acts.

The concept of 'quality regulation', the idea of a regulation capable, among other things, of clearly identifying policy goals and being effective in achieving them, involves both the process that leads to the adoption of norms and their results. With reference to the rule-making process, three different instruments are relevant in our perspective: the *regulatory impact analysis, citizens' involvement* and the *ex post evaluation*.

*Regulatory Impact Analysis* (RIA) is a methodology aiming to evaluate the potential impact of a given piece of legislation before its adoption. The RIA translates into a socio-economic assessment aiming to compare the effect on citizens, enterprises and public administrations potentially deriving from different hypotheses of regulatory intervention. The analysis aims to guide policy-makers to the adoption of more efficient and effective choices while making more explicit the reasons for the choices made. Based on qualitative analysis and, where possible, on the collection of quantitative data, the RIA is also a tool to ensure the coherence and integration of policies, as it requires decision-makers to assess the impact of sectorial policies in other contexts.

*Citizens' involvement* in the design of public policy is typically achieved through consultations, in particular through the *notice and comments* mechanism that consists in seeking public comments on the specific issues to be settled. The collection and the analysis of information about the way in which a regulation is perceived by the citizens helps governments to refine policies, so as to address problems deemed most important by the community and to better define the priorities (*i.e.*, focusing on the areas in which the regulatory intervention is more urgent or, conversely, identifying areas in which the adjustment could be unnecessarily burdensome).

For the *ex post evaluation*, the assessment of regulatory measures already in place, an evaluation is made considering the effects, the impact on the relevant needs and the resources employed. The evaluation produces information that is essential to plan, design and implement new policies, as it allows one to answer essential questions about the relevance, the effectiveness and the cost-benefit ratio of the envisaged rules.

Scientific and methodological results achieved in the CSS research area appear to be supporting the effectiveness of the above-mentioned policy-making instruments mainly in three ways: providing new scientific knowledge about the social phenomena affecting (and affected by) the policy choices; extracting relevant and useful information from the huge amount of digital data available today; offering tools and predictive techniques for the *what-if* analysis.

*RIA* and *ex post evaluation* can find useful support in simulation models. The statistical/ mathematical simulations enabled by *system dynamics* and *microsimulation* models have been already used in the past (the first experiences date back to the 1950s) with reference to specific policy issues, like the prediction of the effects of changes in tax laws on government finances [60]. As we will highlight in the next section, a more powerful tool, in this perspective, is however represented by agent-based simulation models that are particularly suited to study social dynamics strongly influenced by the interactions between agents and other factors (individual motivations, cultural mechanisms, cognitive processes) that are difficult to treat in exclusively mathematical-statistical terms.

As to citizens' involvement, CSS methods seem to be potentially useful to overcome the limits affecting today e-consultation, an activity that is intrinsically connected with the problem of extracting meaningful knowledge from citizens' contributions. Online discussions and feedback contain valuable opinions about the effects of policy decisions and an essential knowledge about the societal needs that the policy tries to address. This information asset is today exploited only to a limited extent by policy-makers due to the variety and the growing amount of information available. Alongside the first web-based e-consultation tools, accepting only textual contributions (e.g., online voting and survey or discussion forums), there are more recent solutions allowing increasingly advanced contributions that span from posts on social media to comments on YouTube videos [61–64].

The problem is that while a limited number of textual contributions can be effectively interpreted by means of statistical methods, the same is not true when the information grows and is spread over different media. In this case, it becomes ever more difficult if not impossible to extract crucial information like the attitudes of the citizens and the main issues they raise (e.g., strengths and weaknesses of the policies under discussion; the effects of implemented decisions; suggestions about potential improvements). This not only results in a loss of knowledge that would be extremely useful to design more contextualized and accepted policies, but prevents from giving citizens automatic feedback about the way in which their opinions have been taken into account, a choice that would help promote government openness and accountability, avoiding disappointment and mistrust in e-consultation processes.

It is no accident that the first OECD report about e-consultation [58] claims that one of the most relevant challenges for the application of the electronic participation model is the automated and cost-efficient extraction of information from the huge amount of unstructured information contained in citizens' contributions. The report explicitly mentions the use of "appropriate technologies" and all of their potential uses: support "the summarization and content analysis of contributions'; help highlighting areas of agreement and disagreement"; "identify the participants main concerns, their level of support for any draft proposals, or their suggestions" [58].

Information extraction research so far has generated several opinion mining methods variously tested in the commercial domain to analyze judgements and reviews posted on the web by customers. These methods today may produce innovative solutions to assess citizens' sentiments (positive, negative or neutral) towards the policies and also to extract the main issues they raise. *Sentiment analysis* [64] is a particularly promising technique in this perspective. The application of computational techniques to determine the attitude of the author of a text (an entire document or even single sentences within it) with respect to a particular topic or, more generally, his or her state of mind is going to offer more insightful knowledge compared to that achievable through the methods that are more traditionally used to evaluate the perception of policies (e.g., the surveys based on the Worldwide Governance Indicators defined by the World Bank to measure the perception of citizens, companies and experts about the quality of governance using six distinct dimensions).

The reasons of the interest toward computational approaches to policy-making should be now clearer. CSS methods could enhance the tools supporting the quality of regulations with positive effects both on the implementation of the legal principles that govern the activity of policy-makers (transparency, accountability, legality, proportionality) and on the overall effectiveness of public policies.

## 5. *In Silico* Management of Social Complexity: Agent-Based Models and Policy Design

Among the different CSS methodologies so far tested in policy-making, social simulation occupies a very special place. The opportunity, offered by the simulations, to better understand and, above all, to somehow predict the evolution of complex socio-economic phenomena has long attracted the attention of scholars with different scientific backgrounds for various reasons who are interested in policy issues.

In an frequently cited essay published in 1998 [65], the American economist Steven Durlauf provides an effective representation of the reasons that should prompt policymakers to look at the scientific paradigm and the investigative methods (simulations in particular) offered by computational social science: social systems and economic dynamics show all of the features of complex phenomena [5,66]. Generated by feedback loops, evolutionary processes and interactions between individuals who constantly change their behavior, the social and economic phenomena are complex, nonlinear and difficult to predict through the most common methods of analysis. However useful, the scientific tools usually supporting policy modeling, mainly based on predictive statistical analyses, show inherent limitations because of their inability to account for the role that the individual and the local dimension plays in determining the overall development of social dynamics. The causal role played, in the evolution of social phenomena, by the individual preferences, by psychological, cognitive and cultural factors or, again, by the reactions of different social actors to political decisions is a puzzle that is difficult to solve with traditional approaches.

The circumstance is confirmed, with specific reference to economic policy, in a statement made by the former European Central Bank president Jean-Claude Trichet in November 2010:

> "When the crisis came, macro models failed to predict the crisis and [...] to explain what was happening [...] in the face of crisis, we felt abandoned by conventional tools [...] The key lesson [...] is the danger of relying on a single tool, methodology or paradigm."

What is needed, in this scenario, is a research methodology allowing one to capture the microfoundations of macro-social outcomes, and the above-mentioned agent-based models represent the most promising solution to this end. Trichet's speech contains a meaningful statement in this regard:

> "A large number of aspects of the observed behavior of financial markets is hard to reconcile with the efficient market hypothesis. But a determinedly empirical approach—Which places a premium on inductive reasoning based on the data, rather than deductive reasoning grounded in abstract

*premises or assumptions—lies at the heart of These methods [...] simulations will play a helpful role.
Agent-based modeling [...] Allows for more complex interactions between agents."*

Having been designed to illuminate and predict the micro-level processes at the root of social macro-phenomena, agent-based models were soon considered as a possible tool to support policy-makers. The focus on this research perspective has grown considerably over the last ten years [67–69], transcending the boundaries of the scientific world and, as shown by the Trichet's speech, reaching the world of institutions.

The issues addressed by the research conducted in this area are very heterogeneous, ranging from the management of environmental resources to the impact of land use decisions, from the effects of economic policies to that of retirements [70–72]. Following the classification proposed in [73] (see, also, for a more detailed analysis, [74]), social simulation models conceived for policy design can be divided into two categories characterized by different theoretical premises and development methods: the *prescriptive* and the *participatory models*, two categories on which it is interesting to offer some detail also in view of the considerations that will be proposed later on.

*Prescriptive models* are inspired by the idea that the scientific explanation of the mechanisms underlying social phenomena can be somehow translated into recommendations for policy-makers. The goal of the researcher (somewhat analogous to that pursued in the legislative sphere through the above-mentioned *Regulatory Impact Analysis*) is to develop a preliminary analysis allowing the policy-makers to understand the phenomenon object of intervention and to make an *ex ante* evaluation of the effects potentially deriving from different policy choices. The core of the research, in this kind of model, consists of theories and scientific evidence related to the target phenomena, elements that are taken into account to create models showing the implications of different choices and to evaluate different solutions before the choice takes place. Studies related to this kind of approach are numerous.

An emblematic and particularly interesting example of this kind of research (because of its connection with public decision-making issues) is the work published by Rouchier and Thoyer in 2003 [75]. The authors use a multi-agent model to study the effects of new decision-making procedures introduced by the EU legislator in the field of genetically-modified organisms, procedures characterized, compared to the previous legislation, by a wider use of public consultation. The aim, in more detail, is to check the impact of the reform on two different aspects of the investigated phenomenon: on the one hand, the level of the influence on political choices deriving from pressure groups and from organizations representing collective interests; secondly, the acceptability to the public of new rules for the regulation of such a delicate matter.

Another interesting example of prescriptive model [76] simulates the negotiation process that, according to the French law, needs to be activated to regulate the use of water within river basins. The model, designed taking into account the different categories of stakeholders involved (farmers, advocacy groups, payers, water companies) and all of the variables relevant for trading (irrigation quotas, price of the water, size of available dams, *etc.*), allows one to estimate the benefits related to different solutions of the negotiation process and to assess the overall impact of the negotiation on the final results of the procedure considered. The purpose is typical of prescriptive models: to provide an information framework to stakeholders in order to improve the quality of policy choices.

*Participatory models.* In this second category of simulations, the development of the model is not only the result of theoretical assumptions or scientific evidences, but derives from a collaborative process in which the recipients of the policies under scrutiny are directly involved. The predictive purposes and decision support are added, in this case, for two other objectives: on the one hand, to achieve a greater degree of adherence of the model to reality; on the other, the beginning of a participatory process and the relevant trading plan for the legitimacy of the choices [77,78]. In participatory modeling process, decision-makers and stakeholders share information and integrate scientific knowledge with the representation of the concrete interests at stake. The models are designed based on observation and developed through an empirical validation process in which the stakeholders are involved directly in the definition of the theme of research and the evaluation of the results. The benefits of participatory methods are different: the acquisition of otherwise inaccessible knowledge; the ability to identify research issues of important practical significance; and above all, a greater likelihood that stakeholders will be sufficiently motivated in the future to implement in their ordinary activity the solutions emerging from the research.

After an initial phase in which policy recipients were involved only for the *ex post* validation of the model, we went, thanks to the spread of online collaboration tools, to more active forms of participation. An interesting example is represented by the *CORMAS* (COmmon-pool Resources and Multi-Agent Simulations) simulation platform developed at the CIRAD (Centre de coopération internationale en recherche agronomique pour le développement), the French research center for the development of agriculture [79] to allow the implementation of participatory simulations intended to support the design and the study of policies relating to the environment and agriculture. To encourage user involvement, participatory simulation exploits different techniques: on the one hand, it uses approaches emerging from the social sciences since the advent of the web (extraction of information from social networks and, more generally, from all of the interactions mediated by the web, online questionnaires, *etc.*); on the other hand, it uses online games as a strategic tool to extract from end users knowledge and information about their needs [80]. Even in the case of participatory models, the experiences to be mentioned are already numerous. Studies conducted so far are mainly linked to sustainable development issues in which the interactions between environmental dynamics, policies, production factors and other economic and social aspects generate complexity that is often difficult to govern [81,82].

A particularly interesting experience [83] is related to the implementation of a multi-agent system developed to stimulate the development, the discussion and comparison of different land management strategies in the Causse Mejean, a limestone plateau in southern France characterized by a delicate ecosystem of grasslands and threatened by an invasion of pines. To allow the identification and comparative assessment of different strategic decisions concerning the location of farms and forest management, the researchers translated the different options into different simulation models defined step-by-step with the direct participation of stakeholders and the use of detailed empirical data concerning the agents involved and the characteristics of the environment.

In the initial phase of the project, the key stakeholders (foresters, farmers, rangers of Cevennes National Park) were invited to contribute to the design of the model through a statement of their views and the proposition of different natural resource management scenarios. Once the model was implemented, the execution of the simulations allowed to assess the impact of each scenario on the production plan (number of sheep to rear, wood growth), on the environment (species in danger of extinction, landscape protection), as well as on other considered relevant territorial entities (forests, farms, grasslands, *etc.*).

The different options have been examined together by all participants in the project passing through the implementation of new scenarios postulated during the discussion. The iteration of the procedure facilitated the identification of a set of compromise solutions based on a shared management of the forests of pines and innovative farming practices protecting the ecosystem.

## 6. Conclusions

In one of his most famous aphorisms Francis Bacon declares: *"human knowledge and human power meet in one; for where the cause is not known the effect cannot be produced. Nature to be commanded must be obeyed"* [84]: in order to master nature (and take advantage of it), we must understand it before. Almost four centuries later, the intuition of the English philosopher, scientist and jurist is more topical than ever, even when applied to the policy-making domain.

Virtually every relevant phenomenon for the decision-maker, from the dynamics of the labor market to the financial crises, from international outbreaks to migratory flows, poses the urgency of a better understanding of real world dynamics. To be more effective and properly contextualized, policy-making need to be increasingly rooted in science. As argued in a growing literature (see, among others, [5,19,68,85–87]), the time is ripe to bridge the gap between the public policies and scientific-methodological advancements that are illuminating social phenomena and the individual processes underlying them.

Mixing social theory and computation, data and laboratory experiments in an innovative way, the computational social science paradigm can contribute to a clearer vision of social processes and, therefore, to the quality of public choices integrating the more traditional quantitative approaches already practiced in social science. However important, statistical analysis is unable by itself to account for the generative processes at the root of social and economic dynamics: being mainly devoted to descriptive purposes and to the discovery of correlations and regularities rather than to the development of scientific hypotheses and theories, statistics may offer a partial view of reality that may well be supplemented by analytical computational social science methods.

As a matter of fact, the computational paradigm is pushing forward our ability to deal with social complexity. It not only takes advantage of unprecedented amounts of ICT data about potentially any kind of human interaction, but is also exploiting computation to inspire, formalize and implement new and more complex theories about real socio-economic systems. It is on the back of these features that CSS promises to turn social science into applicable tools that can inform decision-makers about issues of major concern [19]. New policy analytics could considerably benefit from advanced ICT data mining and analyses, tools that are going to become ever more essential to transform policy in an adaptive, experimental and collaborative process.

Based on the above, the need of a cultural, political and scientific shift becomes clear. If, on the one hand, it is necessary to bring together science and policy design, on the other, it will be essential to promote the cross-fertilization between all of the different cultures, disciplines and research areas involved in the computational social science endeavor, from computer and complexity science to law and economics, from sociology to cognitive and behavioral sciences. In the same perspective, it will be necessary to foster an issue-oriented and interdisciplinary approach to research overcoming a cultural resistance that still appears far from being broken down. As already highlighted, computational social science shares with other emerging inter-disciplinary research areas like cognitive science, the need to develop a paradigm for training new scholars [17]. Universities, research institutes and editorial boards need to understand and reward the effort to work across disciplines.

The stakes are high and deserve major effort: the possibility to successfully cope with the challenges of the contemporary world depends to a significant extent on the capacity to promote a new paradigm for the study of social phenomena and, through it, new practices for policy design. Ten years after Bollinger's talk, the way to go to establish a more strict relationship between science and policy-making is still far from over. Computational social science certainly is not the only solution to the problem, but, for sure, is an important part of our future.

**Acknowledgments:** The author is truly thankful to Margherita Vestoso for the help given in proofreading of the article.

**Conflicts of Interest:** The author declares no conflict of interest.

## References

1. Columbia University Website. Available online: http://www.columbia.edu/node/8184.html (accessed on 29 February 2016).
2. Helbing, D. Globally networked risks and how to respond. *Nature* **2013**, *497*, 51–59. [CrossRef] [PubMed]
3. Miorandi, D.; MALTESE, V.; Rovatsos, M.; Nijholt, A.; Stewart, J. *Social Collective Intelligence: Combining the Powers of Humans and Machines to Build a Smarter Society*; Springer: New York, NY, USA, 2014.
4. Walker, B.; Salt, D. *Resilience Thinking: Sustainable Ecosystems and People in a Changing World*; Island Press: Washington, DC, USA, 2006.
5. Squazzoni, F. A social science-inspired complexity policy: Beyond the mantra of incentivization. *Complexity* **2014**, *19*, 5–13. [CrossRef]
6. Ormerod, P. *N Squared: Public Policy and the Power of Networks*; RSA Pamphlets: London, UK, 2010.

7.   Coleman, J.S.; Fararo, T.J. *Rational Choice Theory*; Sage: New York, NY, USA, 1992.
8.   Kahneman, D. Maps of bounded rationality: Psychology for behavioral economics. *Am. Econ.* **2003**, *93*, 1449–1475. [CrossRef]
9.   Smith, V. Constructivist and ecological rationality in economics. *Am. Econ.* **2003**, *93*, 465–508. [CrossRef]
10.  Anderson, C. The End of Theory: The Data Deluge Makes the Scientific Method Obsolete. *Wired Mag.* **2008**. Available online: http://www.wired.com/science/discoveries/magazine/16-07/pb_theory (accessed on 25 February 2016).
11.  Cioffi-Revilla, C. Computational social science. *Comput. Stat.* **2010**, *2*, 259–271. [CrossRef]
12.  Mayer-Schönberger, V.; Cukier, K. *Big Data: A Revolution that Will Transform How We Live, Work, and Think*; Houghton Mifflin Harcourt: Boston, MA, USA, 2013.
13.  Ayres, I. *Super Crunchers: Why Thinking-by-Numbers is the New Way to Be Smart*; Bantam Dell: New York, NY, USA, 2007.
14.  Boulton, G.; Campbell, P.; Collins, B.; Elias, P.; Hall, W.; Laurie, G.; Walport, M. *Science as an Open Enterprise*; Royal Society: London, UK, 2012; p. 104.
15.  Tansley, S.; Tolle, K. *The Fourth Paradigm: Data-Intensive Scientific Discovery*; Microsoft research: Redmond, WA, USA, 2009.
16.  Bell, G.; Hey, T.; Szalay, A. Beyond the Data Deluge. *Science* **2009**, *323*, 1297–1298. [CrossRef] [PubMed]
17.  Lazer, D.; Pentland, A.; Adamic, L.; Aral, S.; Barabási, A.; Brewer, D.; Christakis, N.; Contractor, N.; Fowler, J.; Gutmann, M. Computational Social Science. *Science* **2009**, *323*, 721–723. [CrossRef] [PubMed]
18.  Vespignani, A. Predicting the behavior of techno-social systems. *Science* **2009**, *325*, 425–428. [CrossRef] [PubMed]
19.  Conte, R.; Gilbert, N.; Bonelli, G.; Cioffi-Revilla, C.; Deffuant, G.; Kertesz, J.; Loreto, V.; Moat, S.; Nadal, J.-P.; Sanchez, A.; *et al.* Manifesto of computational social science. *Eur. Phys. J. Spec. Top.* **2012**, *214*, 325–346. [CrossRef]
20.  Cardie, C.; Wilkerson, J. Text Annotation for Political Science Research. *J. Inf. Technol. Politics* **2008**, *5*, 1–6. [CrossRef]
21.  Kaal, B.; Maks, I.; Van Elfrinkhof, A. *From Text to Political Positions: Text Analysis Across Disciplines*; John Benjamins Publishing Company: Amsterdam, The Netherlands, 2014.
22.  Lewis, K.; Kaufmana, J.; Gonzaleza, M.; Wimmerb, A.; Christakisa, N. Tastes, ties, and time: A new social network dataset using Facebook.com. *Soc. Netw.* **2008**, *30*, 330–342. [CrossRef]
23.  Onnela, J.P.; Saramaki, J.; Hyvonen, J.; Szabo, G.; Lazer, D.; Kaski, K.; Kertesz, J.; Barabasi, A.-L. Structure and tie strengths in mobile communication networks. *Proc. Natl. Acad. Sci. USA* **2007**, *104*, 7332–7336. [CrossRef] [PubMed]
24.  Balcan, D.; Colizza, V.; Gonçalves, B.; Hu, H.; Ramasco, J.J.; Vespignani, A. Multiscale mobility networks and the spatial spreading of infectious diseases. *Proc. Natl. Acad. Sci. USA* **2009**, *106*, 21484–21489. [CrossRef] [PubMed]
25.  Karsai, M.; Kivelä, M.; Pan, R.K.; Kaski, K.; Kertész, J.; Barabási, A.-L.; Saramäki, J. Small but slow world: How network topology and burstiness slow down spreading. *Phys. Rev. E* **2011**, *83*, 025102. [CrossRef] [PubMed]
26.  Squazzoni, F. *Epistemological Aspects of Computer Simulation in the Social Sciences*; Springer: New York, NY, USA, 2009.
27.  Anzola, D. The Philosophy of Computational Social Science. Ph.D. Thesis, University of Surrey, Guildford, UK, 2015.
28.  Cioffi-Revilla, C. *Introduction to Computational Social Science: Principles and Applications*; Springer: Berlin, Germany, 2013.
29.  Kerren, A.; Stasko, J.; Fekete, J.; North, C. *Information Visualization: Human-Centered Issues and Perspectives*; Springer: Berlin, Germany, 2008.
30.  Keim, D. *Visual Analytics: Scope and Challenges*; Springer Berlin: Heidelberg, Germany, 2008.
31.  Hermann, T.; Hunt, A. *The Sonification Handbook*; Logos Verlag: Berlin, Germany, 2011.
32.  Heise, D. *Understanding Events: Affect and the Construction of Social Action*; Cambridge University Press: New York, NY, USA, 1979.
33.  Aggarwal, C.; Zhai, C. *Mining Text Data*; Springer: New York, NY, USA, 2012.

34. Lettieri, N.; Malandrino, D.; Spinelli, R. Text and Social Network Analysis As Investigative Tools: A Case Study. In *Law and Computational Social Science*; Lettieri, N., Faro, S., Eds.; ESI: Naples, Italy, 2013.

35. Auyang, S. *Foundations of Complex-System Theories in Economics, Evolutionary Biology, and Statistical Physics*; Cambridge University Press: Cambridge, UK, 1998.

36. Gros, C. *Complex and Adaptive Dynamical Systems: A Primer*; Springer: New York, NY, USA, 2008.

37. Mitchell, M. *Complexity: A Guided Tour*; Oxford University Press: Oxford, UK, 2009.

38. Miller, J.; Page, S. *Complex Adaptive Systems: An Introduction to Computational Models of Social Life*; Princeton University Press: Princeton, NJ, USA; Oxford, MS, USA, 2009.

39. Lux, T. Financial Power Laws: Empirical Evidence, Models, Mechanisms. In Proceedings of the International Workshop on Power Laws in the Social Sciences, Center for Social Complexity, George Mason University, Fairfax, VA, USA, 10–13 July 2007.

40. Kellstedt, P.M. Race Prejudice and Power Laws of Extremism. In Proceedings of the International Workshop on Power Laws in the Social Sciences, Center for Social Complexity, George Mason University, Fairfax, VA, USA, 10–13 July 2007.

41. Scott, J. *Social Network Analysis*; Sage: New York, NY, USA, 2012.

42. Milroy, L. Social network analysis and language change: Introduction. *Eur. J. Engl. Stud.* **2000**, *4*, 217–223. [CrossRef]

43. Solé, R.V.; Corominas-Murtra1, B.; Valverde1, S.; Steels, L. Language networks: Their structure, function, and evolution. *Complexity* **2010**, *15*, 20–26.

44. Berry, B.; Griffth, D.; Tiefelsdorf, M. From Spatial Analysis to Geospatial Science. *Geograph. Anal.* **2008**, *40*, 229–238. [CrossRef]

45. Sterman, J. *Business Dynamics: System Thinking and Modeling for a Complex World*; Irwin McGraw-Hill Companies: Boston, MA, USA, 2000.

46. Gilbert, N.; Troitzsch, K. *Simulation for the Social Scientist*, 3rd ed.; Open University Press: Philadelphia, PA, USA, 2008.

47. Epstein, J.M. *Generative Social Science: Studies in Agent-Based Computational Modeling*; Princeton University Press: Princeton, NJ, USA, 2006.

48. Cioffi-Revilla, C.; Rouleau, M. MASON RebeLand: An Agent-Based Model of Politics, Environment, and Insurgency. *Int. Stud. Rev.* **2010**, *12*, 31–52. [CrossRef]

49. EU Commission, Directorate-General for Research and Innovation Socio-Economic Sciences and Humanities. Scientific Evidence for Policy-Making: Research Insights from Socio-Economic Sciences and Humanities. Available online: http://goo.gl/FwS9sY (accessed on 25 February 2016).

50. FuturICT Project. Available online: http://www.futurict.eu (accessed on 25 February 2016).

51. FOC Project. Available online: http://www.focproject.eu (accessed on 25 February 2016).

52. EURACE Project. Available online: http://eurace.org (accessed on 25 February 2016).

53. FUPOL Project. Available online: http://www.fupol.eu/en (accessed on 25 February 2016).

54. Battiston, S.; Caldarelli, G.; Georg, C.; May, R.; Stiglitz, J. Complex derivatives. *Nat. Phys.* **2013**, *9*, 123–125. [CrossRef]

55. Sonntagbauer, S.; Sonntagbauer, P.; Nazemi, K.; Burkhardt, D. The FUPOL policy lifecycle. In *Handbook of Research on Advanced ICT Integration for Governance and Policy Modeling*; IGI Global: Hershey, PA, USA, 2014; pp. 61–87.

56. Althaus, C.; Bridgman, P.; Davis, G. *The Australian Policy Handbook*; Allen & Unwin: Crows Nest, Sydney, Australia, 2012.

57. Lettieri, N.; Faro, S. Computational social science and its potential impact upon law. *Eur. J. Law Technol.* **2012**, *3*. Available online: http://ejlt.org/article/view/175 (accessed on 25 February 2016).

58. OECD. Recommendation of the Council on Regulatory Policy and Governance. Available online: http://www.oecd.org/dataoecd/45/55/49990817.pdf (accessed on 25 February 2016).

59. EU Commission. *European Governance. A White Paper*. Available online: http://goo.gl/im749B (accessed on 25 February 2016).

60. Troitzsch, K. Legislation, Regulatory Impact Assessment and Simulation. In *Law and Computational Social Science*; Lettieri, N., Faro, S., Eds.; ESI: Naples, Italy, 2013; pp. 57–75.

61. Charalabidis, Y.; Gionis, G.; Ferro, E.; Loukis, E. Towards a Systematic Exploitation of Web 2.0 and Simulation Modeling Tools in Public Policy Process. In *Electronic Participation*; Springer: Berlin, Germany; Heidelberg, Germany, 2010; pp. 1–12.

62. Maragoudakis, M.; Loukis, E.; Charalabidis, Y. A review of opinion mining methods for analyzing citizens' contributions in public policy debate. In *Electronic Participation*; Springer: Berlin, Germany; Heidelberg, Germany, 2011; pp. 298–313.

63. Mola, L.; Pennarola, F.; Sa, S. *From information to smart society: Environment, politics and economics*; Springer: New York, NY, USA, 2014.

64. Rose, J.; Sanford, C. Mapping eParticipation Research: Four Central Challenges. *Commun. Assoc. Inf. Syst.* **2007**, *20*, 909–943.

65. Durlauf, S. What should policymakers know about economic complexity? *Wash. Q.* **1998**, *21*, 155–165. [CrossRef]

66. Ahrweiler, P. *Innovation in Complex Social Systems*; Routledge: London, UK, 2010.

67. Buchanan, M. Meltdown Modeling. Could Agent-Based Computer Models Prevent Another Financial Crisis? *Nature* **2009**, *460*, 680–682. [CrossRef] [PubMed]

68. Farmer, J.; Foley, D. The economy needs agent-based modelling. *Nature* **2009**, *460*, 685–687. [CrossRef] [PubMed]

69. Helbing, D. *Social Self-Organization*; Springer: New York, NY, USA, 2012.

70. Ma, Y.; Zhenjiang, S.; Kawakami, M. Agent-Based Simulation of Residential Promoting Policy Effects on Downtown Revitalization. *J. Artif. Soc. Soc. Simul.* **2013**, *16*. Available online: http://jasss.soc.surrey.ac.uk/16/2/2.html (accessed on 25 February 2016). [CrossRef]

71. Saam, N.J.; Kerber, W. Policy Innovation, Decentralised Experimentation, and Laboratory Federalism. *J. Artif. Soc. Soc. Simul.* **2013**, *16*. Available online: http://jasss.soc.surrey.ac.uk/16/1/7.html (accessed on 25 February 2016). [CrossRef]

72. Werker, T.; Werker, C. Policy Advice Derived from Simulation Models. *J. Artif. Soc. Soc. Simul.* **2009**, *12*. Available online: http://jasss.soc.surrey.ac.uk/12/4/2.html (accessed on 25 February 2016).

73. Squazzoni, F.; Boero, R. Complexity-friendly policy modelling. In *Innovation in Complex Systems*; Ahrweiler, P., Ed.; Routledge: London, UK, 2010.

74. Yücel, G.; Van Daalen, E. An objective-based perspective on assessment of model-supported policy processes. *J. Artif. Soc. Soc. Simul.* **2009**, *12*. Available online: http://jasss.soc.surrey.ac.uk/12/4/3.html (accessed on 25 February 2016).

75. Rouchier, J.; Thoyer, S. Modelling a European decision making process with heterogeneous public opinion and lobbying: The case of the authorization procedure for placing genetically modified organisms on the market. In *Multi-Agent-Based Simulation III, Lecture Notes in Computer Science*; Hales, D., Edmonds, B., Norling, E., Rouchier, J., Eds.; Springer: Berlin, Germany, 2003; p. 149.

76. Thoyer, S.; Morardet, S.; Rio, P.; Simon, L.; Goodhue, R.; Rausser, G. A Bargaining Model to Simulate Negotiations between Water Users. *J. Artif. Soc. Soc. Simul.* **2001**, *4*. Available online: http://jasss.soc.surrey.ac.uk/4/2/6.html (accessed on 25 February 2016).

77. Ramanath, A.; Gilbert, N. The design of participatory agent-based social simulations. *J. Artif. Soc. Soc. Simul.* **2004**, *7*. Available online: http://jasss.soc.surrey.ac.uk/7/4/1.html (accessed on 25 February 2016).

78. Nguyen-Duc, M.; Drogoul, A. Using Computational Agents to Design Participatory Social Simulations. *J. Artif. Soc. Soc. Simul.* **2007**, *10*. Available online: http://jasss.soc.surrey.ac.uk/10/4/5.html (accessed on 25 February 2016).

79. Le Page, C.; Becu, N.; Bommel, P.; Bousquet, F. Participatory Agent-Based Simulation for Renewable Resource Management: The Role of the Cormas Simulation Platform to Nurture a Community of Practice. *J. Artif. Soc. Soc. Simul.* **2012**, *15*. Available online: http://jasss.soc.surrey.ac.uk/15/1/10.html (accessed on 25 February 2016). [CrossRef]

80. Gilbert, N.; Maltby, S.; Asakawa, T. Participatory simulations for developing scenarios in environmental resource management. In *Third Workshop on Agent-Based Simulation*; Urban, C., Ed.; SCS European Publishing House: Passau, Germany, 2002; p. 67.

81. Parker, D.C.; Manson, S.M.; Janssen, M.A.; Hoffmann, M.J.; Deadman, P. Multi-agent systems for the simulation of land-use and land-cover change: A review. *Ann. Assoc. Am. Geogr.* **2003**, *93*, 314–337. [CrossRef]

82. Barreteau, O.; Bousquet, F.; Millier, C.; Weber, J. Suitability of Multi-Agent Simulations to Study Irrigated System Viability: Application to Case Studies in the Senegal River Valley. *Agric. Syst.* **2004**, *80*, 255–275. [CrossRef]

83. Etienne, M.; Le Page, C.; Cohen, M. A Step-By-Step Approach to Building Land Management Scenarios Based on Multiple Viewpoints on Multi-Agent System Simulation. *J. Artif. Soc. Soc. Simul.* **2003**, *6*. Available online: http://jasss.soc.surrey.ac.uk/6/2/2.html (accessed on 25 February 2016).

84. Bacon, F. *Novum Organum*; P.F. Collier: New York, NY, USA, 1902; Available online: http://oll.libertyfund.org/titles/1432 (accessed on 25 February 2016).

85. Ormerod, P. Networks and the need for a new approach to policymaking. In *Complex New World. Translating New Economic Thinking into Public Policy*; Dolphin, T., Nash, D., Eds.; Institute for Policy Research: London, UK, 2012; pp. 28–38.

86. Lane, D.; Pumain, D.; van der Leeuw, S. *Complexity Perspectives in Innovation and Social Change*; Springer Verlag: Berlin, Germany, 2009.

87. Prewitt, K.; Schwandt, T.A.; Straf, M. *Using Science as Evidence in Public Policy*; National Academies Press: Washington, DC, USA, 2012; Available online: http://biblioteca.ucv.cl/site/colecciones/manuales_u/13460.pdf (accessed on 25 February 2016).

*future internet*

MDPI

*Article*

# Case Study: IBM Watson Analytics Cloud Platform as Analytics-as-a-Service System for Heart Failure Early Detection

**Gabriele Guidi, Roberto Miniati, Matteo Mazzola and Ernesto Iadanza ***

Department of Information Engineering Unversità degli Studi di Firenze, v. S. Marta, 3-50139 Firenze, Italy; gabriele.guidi@unifi.it (G.G.); roberto.miniati@gmail.com (R.M.); matteo.mazzola@stud.unifi.it (M.M.)
* Correspondence: ernesto.iadanza@unifi.it; Tel.: +39-347-592-2874

Academic Editor: Dino Giuli
Received: 14 February 2016; Accepted: 24 June 2016; Published: 13 July 2016

**Abstract:** In the recent years the progress in technology and the increasing availability of fast connections have produced a migration of functionalities in Information Technologies services, from static servers to distributed technologies. This article describes the main tools available on the market to perform Analytics as a Service (AaaS) using a cloud platform. It is also described a use case of IBM Watson Analytics, a cloud system for data analytics, applied to the following research scope: detecting the presence or absence of Heart Failure disease using nothing more than the electrocardiographic signal, in particular through the analysis of Heart Rate Variability. The obtained results are comparable with those coming from the literature, in terms of accuracy and predictive power. Advantages and drawbacks of cloud versus static approaches are discussed in the last sections.

**Keywords:** cloud; decision support systems; data mining; Heart Failure

---

## 1. Introduction

In the recent years the progress in technology and the increasing availability of fast connections has produced a migration of functionalities in Information Technology (IT) services, from static servers to distributed technologies. This phenomenon is commonly well known as *Cloud Computing*; the most exhaustive and official definition comes from the US National Institute of Standards and Technology (NIST) [1], which introduces all the fundamental concepts of the cloud systems, such as on-demand access to resources by the end user and offering services with minimal infrastructures and management effort.

NIST definition points out that cloud computing includes data processing and data storage, both performed on remote servers.

The arrival of cloud computing is also changing many core concepts in IT, defining new service models for distribution to final customers. Summarizing the definitions in [1]:

- *Software as a Service (SaaS)*: the consumer can use various cloud devices to take advantage of a provider's application (web application) that is stored on a cloud infrastructure.
- *Platform as a Service (PaaS)*: business users can deploy and distribute their applications onto the cloud, taking advantage of the tools supported by the provider without having to manage the underlying infrastructure
- *Infrastructure as a Service (IaaS)*: in addition to all the functionalities of the PaaS model, the user can also control the operating system and the storage as well as select some network components (e.g., host firewalls).

We can therefore understand the importance of cloud technology, primarily in business environment. The companies that choose to move their assets on the cloud will only take care of their core-business (e.g., software developing, for a software house). No more need for data management strategies (security, persistence, geographically scattered backups, etc.) nor hardware updating to guarantee adequate computing power and storage space.

The end user, by accessing a SaaS from a common browser, benefits of all the software features without the need for an adequate hardware or installation/configuration steps.

In research, often the set up of a proper work environment is a time consuming and costly activity. In order to guarantee enough storage space and computing power, new hardware has a short obsolescence time. In machine-learning and data-mining, the researchers often deal with High Performance Computing (HPC) and Big Data. These are the typical conditions where cloud technologies can offer the best advantages: large amounts of data and high computational power. Cloud architectures applied to these contests have been called Analytics-as-a-Service (AaaS) and have been largely treated in literature in the last years [2–6].

In this article we will start describing some of the available systems and platforms for analytics and their specificities. Afterwards we will describe a case study where one of these tools has been tested to face a problem concerning a pathology called Hearth Failure (HF), very well described in literature. We used machine-learning to predict the presence of the disease, relying only on the Hearth Rate Variability (HRV) analysis.

The following is a brief summary of the available analytics systems/platforms. We selected the top ten systems according to the rankings from Martin Butler [7], and basing on the review published by Butler Analytics [8].

## 2. Related Studies: State of the Art about Analytics Tools on the Market

The following information has been extracted from the public websites of vendors and from the above mentioned reviews by M. Butler [7]. The intent is exclusively to provide an overview on the state of art about the currently available cloud-based products for analytics.

### 2.1. Revolution Analytics

Revolution Analytics is a system of analysis based on the R language, which is a well-known programming language widely used for statistical issues. Revolution Analytics integrates this open source language into a form easily usable by enterprises. Revolution R is distributed in two versions, Open and Enterprise editions, and offers high-performance, scalable, enterprise-capable analytics and Big Data analysis. The Enterprise edition supports several ready to use tools that allow users to exploit features for model building/deployment, as well as advanced data analysis. Revolution Analytics was recently acquired by Microsoft, which has promoted a campaign of facilities for academic accounts and non-profit associations.

### 2.2. Statistica

Statistica is a suite from StatSoft, recently acquired by Dell Software. The Statistica suite offers many products, such as Data Visualization, BigData, DataMiner, TextMiner, decision-making and sentiment analysis tools. The application fields where Statistica has specific solutions are multiple: cross-industry, energy oil and gas, financial, healthcare, insurance, manufacturing, pharmaceutical. In the healthcare field, Statistica offers a variety of graphical modules that enable analytics for several tasks like: patient/customers profiling, prediction of hospital readmissions, cost estimation, risks management.

### 2.3. Oracle Advanced Analytics

Oracle Advanced Analytics is an analytics platform based on Oracle Database that includes two tools: Oracle R Enterprise and Oracle Data Mining. The specificity of the Oracle Analytics System is that it is provided by a company whose core business is database. This is reflected in the product Oracle Advanced Analytics in providing data analysis directly on data that are stored in Oracle Database: customers can run the algorithms directly where the data are located, in the database (no slow input-output operations). As mentioned, Oracle offers two types of systems: Oracle R Enterprise that allows users to use their R-language skills and tools to analyze their data, and Oracle Data Mining that allows users to create data mining functions using SQL language.

## 2.4. FICO

FICO provides predictive analytics with the peculiarity of being combined with prescriptive analytics and business rules management. It offers specialized solutions oriented to market, such as functions for customers engagement, or oriented to the bank scoring, as mortgage calculation or risk functions. FICO has an entire section dedicated to analytics on the cloud, providing these technologies to be suitable also for smaller businesses without having to keep local analytics server.

## 2.5. KXEN

KXEN in 2013 was acquired by SAP. KXEN originally offered solutions primarily geared to risk minimization including heavy duty products, mainly suitable for large organizations.

## 2.6. Salford Systems

Salford offers the SPM (Software Predictive Modeler) suite for analytics and data mining. It includes some of the most popular machine learning algorithms, such as Leo Breiman CART—Classification And Regression Trees (Salford offers a patented extended version); Random Forests; MARS—multivariate Adaptive Regression Splines; TreeNet (Stochastic Gradient Boosting). Salford also offers custom demonstrations called Rapid Response Data Mining to evaluate the potential benefits that a company could obtain from its adoption and the ROI (Return Of Investment).

## 2.7. TIBCO Spotfire

TIBCO Spotfire provides analytics capabilities focused on the management of events (Complex Event Processing) and data visualization. The product is offered in three variants: Desktop (for single local users), Cloud (SaaS) and Platform (business oriented scalable solution), all including visual tools and advanced analytics. In addition, TIBCO offers specific vertical solutions for the energy field, finance, manufacturing, customer management and telecommunications.

## 2.8. SAP Predictive Analytics

SAP is perhaps one of the most famous management software for business, providing both general purpose and special purpose tools in various fields. SAP Predictive Analytics has the advantage that many companies, already relying on SAP systems, may integrate SAP predictive analytics without having to refer to another software company, thus obtaining a full integrated macro-solution.

The SAP product provides predictive analytics automation, Big Data analysis, model management, predicting score and more. An interesting feature is that this product offers a tight integration with R language to give the user the chance of using a large number of available algorithms and reusing custom R scripts.

## 2.9. SAS

SAS (Statistical Analytics System) [9] is a big company that started its activities by offering analytical services for agriculture; today SAS offers services in all application areas, from the academic field, to the life science field, from medical to management-aid. SAS offers a vast set of solutions whose names recall the area of application, such as: SAS Curriculum Pathways, SAS Data Management, SAS Visual Analytics, etc.

Starting from SAS 9.4, the solution is deployable also onto the cloud. To be used with its full power, the system is mainly directed to expert users. The fields of application are several, mainly related to business intelligence and bank trading. We also found examples of applications in healthcare ([10,11]) and Big Data ([12]).

*2.10. IBM Watson Analytics*

IBM Watson Analytics [13] is a cloud based system that allows the final user to run complex analytics using a simple interface, using nothing but a web browser (no specific clients or plug-ins to be installed on local machines). The goal is allowing users, experts who may be familiar with data analytics techniques or not, to focus only on their experiment or case study.

Once the database is uploaded on the cloud, the system offers three categories of functions: Explore, Predict and Assemble (Figure 1). The "Explore" mode provides data clustering to detect patterns and intrinsic relationships between data (non-supervised training techniques). The "Predict" mode allows the user to perform predictions on the data, disclosing the predictive strength of the most significant parameters, compared to a single target parameter set up by the user. "Assemble" mode is dedicated to efficiently show data using infographics.

The fields of application of such general purpose systems are very large; there is a Watson Analytics Community where users can share use cases as samples of application in various areas [14].

For example, the system is used by a human resource manager to identify the parameters that affect the workers resignations [15]. One more example is the analysis of product sales near particular events, such as fireworks near July 4th in US [16]. Other use cases include banking, insurance, retail, telecommunications, government, nonprofit, education, marketing, sales, information technology, finance and more.

One of the strengths of Watson Analytics, is the automation of many steps of the analysis, allowing also non expert users to start using it. Main automation functionalities can be summarized as:

- Automatic Data Preparation

  - Data Transformation
  - Data Quality Index, based on empty fields analysis and constant values identification

- Automatic Modeling

  - Auto selection of best models and detection of strongest relationships: Decision Tree (CHAID) and Key Driver
  - Auto selection of best predictive statistical method basing on data type: Watson Analytics automatically chooses the best regression model for the user data between linear, logistic, multivariate etc.

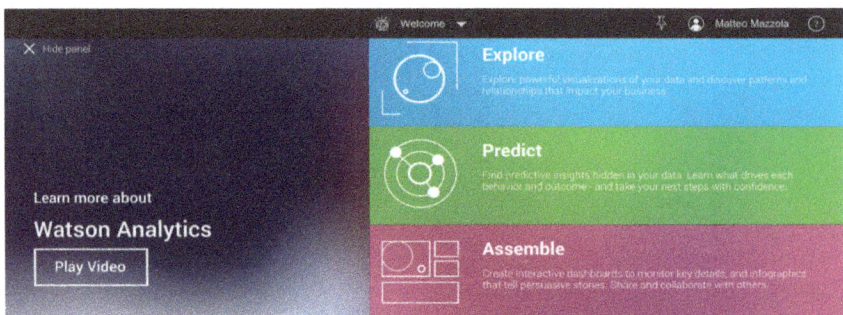

**Figure 1.** Watson Analytics (WA) home screen showing three modalities: Explore, Predict and Assemble.

In addition, Watson Analytics includes an engine for text cognitive analysis (IBM is the world leader); the user can submit a question typed in natural language to extract information from data.

## 3. Use Case: Watson Analytics as AaaS to Identify HF Patients Analyzing Only the ECG Signal

### 3.1. Purpose of the Use Case

The goal of this use case is to report the experience of using a well-known AaaS system, IBM Watson Analytics, for medical information technology research purposes. The chosen task is: identifying the presence of Heart Failure (HF) by the simple analysis of the electrocardiographic signal (ECG). The research trends for analysis and decision support systems in HF are typically performed using two different approaches. One seeks to obtain predictive models that are as exhaustive as possible by including in the analysis a big amount of parameters of different nature (blood tests, biometrical values, etiology, comorbidities etc.) [17–19]. The other approach, through the application of tele-care systems, tries to obtain predictive models by restricting the inputs to easily measurable parameters detectable in tele-monitoring environments [20–22].

This study belongs to the second line of research and has the purpose of determining whether the analysis of the Heart Rate Variability (HRV) on an ECG signal is enough to determine the presence or absence of HF, and if the result with a signal acquired for 24 h (Long Term) is comparable to an acquisition of five minutes (Short Term). The HRV technique will be described later in the article.

The Prediction Strength (PS) obtained using cloud Watson Analytics system will be evaluated as well. The obtained results will be assessed in comparison with the existing literature, where the same goal is pursued by analyzing similar signals with different modalities (analytics using machine learning, non-cloud techniques) [23,24].

### 3.2. Medical and ECG-Analysis Background

#### 3.2.1. Heart Failure

The Heart Failure is an alteration of the structure and function of the heart that involves the body's inability to provide the proper amount of blood to the organism. The body's reaction to HF causes sodium and water accumulation in the lungs and tissues causing fatigue, difficulty to perform physical efforts, shortness of breath and pulmonary edema. As time goes, patient's condition can worsen increasing the severity of lung edema and causing death. Clinical course of the disease leads the patient through a chronic stage that is quite stable but often alternated with worsening requiring hospitalization. In some cases, these severe episodes could be avoided with a preventive therapy. Obtaining additional information on this disease and its evolution would be greatly important for the health of patients. The overall prevalence of HF is slowly increasing due to the aging of the population and the success in the survival of patients suffering from a heart attack [25].

The literature shows improved outcomes for HF patients supported by tele-monitoring systems [26,27]; hence, a non-invasive system of investigation on HF presence based on analysis of ECG recordings (possibly Short-Term), may be useful to timely administer a specific therapy and to prevent worsening.

#### 3.2.2. Heart Failure Diagnosis Methods

The "ESC (European Society of Cardiology) Guidelines for the diagnosis and treatment of acute and chronic heart failure" [25] make explicit that the diagnosis of HF is a complicated process because symptoms are often similar to those of other diseases. In addition, symptoms specifically related to HF (i.e., orthopnoea and paroxysmal nocturnal dyspnoea) are less common. Monitoring of symptoms and signs may be useful to evaluate the effectiveness of the therapy, but for HF diagnosis physicians need to get objective evidence of a structural or functional cardiac abnormality by instrumental examination. The ESC Guidelines clearly state which tests are required for a comprehensive diagnosis:

- Echocardiogram: it provides immediate information about the volumes of the atrial and ventricular chambers, and in particular about the ejection fraction.
- Electrocardiogram: it provides information about the rhythm of the heartbeat and possible faults in the electrical signal transmission (atrioventricular block etc.).
- Natriuretic Peptides: they are another important marker of HF. The examination consists in the analysis of blood concentration of the BNP (B-type Natriuretic Peptide) or NT-proBNP (N-terminal pro-B type), hormones secreted in abnormal amounts when the heart is diseased or the load on any chamber is increased.
- Chest X-ray: this examination is often more useful to identify lung diseases that cause symptoms similar to HF. However, this examination may show pulmonary venous congestion or edema in a patient with HF.
- Other laboratory tests: the guidelines indicate a multitude of laboratory parameters that may be related to HF, including biochemical (sodium, potassium, creatinine) and hematological tests (hemoglobin, hematocrit, ferritin, Leucocytes, and platelets). Also thyroid hormone is important because it can have an impact on HF.

ESC Guidelines also report a complex graph showing the algorithm that manages HF diagnostic decisions (a sequence of given tests related to the symptoms).

In this complex scenario, some of the mentioned useful parameters for HF diagnosis are not appropriate for a home telemonitoring context. One of the aims of this study is to establish in which cases the sole ECG-HRV analysis could be appropriate to perform a preliminary and early diagnosis of HF.

### 3.2.3. ECG Signal Analysis through HRV

The electrocardiogram (ECG) is the graphic reproduction of the heart activity during its cycle of operation, recorded via sensors (electrodes) placed on the skin. Specifically, the cardiac activity can be estimated by measuring the voltage differences in some defined point of the body. For decades the electrocardiogram has been the easiest, practical, less invasive and less expensive method to observe the electrical activity of the heart. The ECG outcome has a characteristic shape, whose variations can indicate problems. It contains several sections called waves, positive and negative, which are repeated for each cardiac cycle:

- P wave: the first wave of the cycle, which corresponds to ventricular depolarization of the atria; the contraction is quite weak and the wave is small.
- QRS complex: set of three waves in rapid succession corresponding to the depolarization of the ventricles: the Q wave is negative and small, the R is a high positive peak, while S is again a small negative wave.
- T Wave: it refers to the ventricle repolarization.
- U Wave: due to the repolarization of the papillary muscles, which is also not always identifiable
- ST Section: period during which the ventricular cells are depolarized, therefore isoelectric, so electrical changes are not greater than 1 mm on the graph.
- QT interval: interval in which occurs ventricular depolarization and repolarization; its duration varies with the heart rate, but generally remains between 350 and 440 ms.

Heart Rate Variability (HRV) nomenclature refers to the physiological phenomenon of time length variation between two heart beats; once defined the peak wave in the cardiac cycle as "R", we can also refer to HRV as "RR variation" or "RR interval", meaning the time frame between two R waves. HRV can be performed using two ways:

- Long-term analysis: performed on a ECG signal acquired for 24 h in a row, using a device called Cardiac Holter

- Short-term analysis: performed on a ECG signal acquired for just 5 min or less

HRV analysis can be carried on in both time and frequency domain. The values obtained from the ECG signal performing the time domain analysis are summarized in [28], and are:

- SDANN: Standard deviation of the averages of NN intervals in all 5-min segments of a 24-h recording
- AVNN: Average of all NN intervals
- pNN50: Percentage of differences between adjacent NN intervals that are greater than 50 ms; a member of the larger pNNx family
- SDNNIDX: Mean of the standard deviations of NN intervals in all 5-min segments of a 24-h recording
- rMSSD: Square root of the mean of the squares of differences between adjacent NN intervals
- SDNN: Standard deviation of all NN intervals

In the frequency domain parameters from the ECG signal are:

- LF/HF: Ratio of low to high frequency power
- TOTPWR: Total spectral power of all NN intervals up to 0.04 Hz
- LF: Total spectral power of all NN intervals between 0.04 and 0.15 Hz.
- ULF: Total spectral power of all NN intervals up to 0.003 Hz
- HF: Total spectral power of all NN intervals between 0.15 and 0.4 Hz
- VLF: Total spectral power of all NN intervals between 0.003 and 0.04 Hz

### 3.3. Material and Methods

The diagram in Figure 2 shows the workflow of the study that has been carried out.

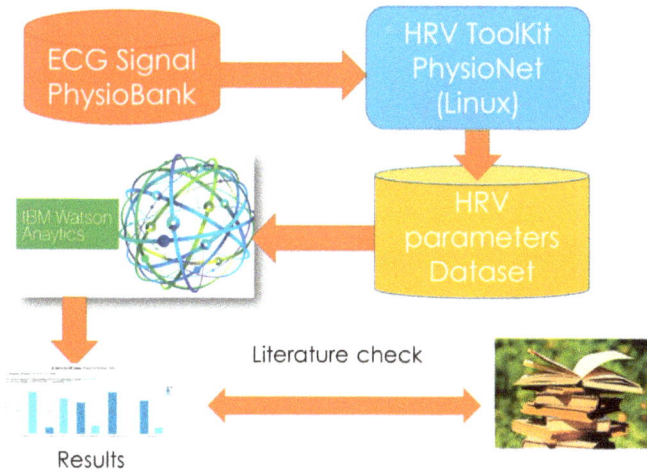

**Figure 2.** Diagram showing the workflow our study.

### 3.3.1. Dataset ECG Signals

The ECG signals on which the analysis has been performed were obtained from the PhysioBank PhysioNet public database [28]. The data used for the test were extracted from three separate datasets found in PhysioBank database:

- CHFDB: Congestive Heart Failure Database contains 15 subjects including 11 men, (age range: 22–71), and 4 women (age range 54–63), with high severity of heart failure disease.

- CHF2DB: contains 29 subjects aged between 34 and 79 years with medium severity of heart failure; the subjects include 8 men and 2 women; the sex of the remaining 19 patients is not known.
- NSR2DB: the Normal Sinus Rhythm Database contains 54 healthy subjects including 30 men (age range: 28–76), and 24 women (age range 58–73).

Table 1 summarizes the overall dataset analyzed.

**Table 1.** Dataset distribution.

| Number of Healthy Patients | Number of HF Patients |
|:---:|:---:|
| 54 | 44 |

### 3.3.2. Extraction of HRV Parameters

For the extraction of HRV parameters we used the tool set provided by PhysioNet called HRV Toolkit, used on Ubuntu Linux.

In order to make repeatable tasks, we report some details on the data extraction, that has been performed creating two scripts, which recall separately the short-term and long-term analysis, both set by literature instructions: 5 min time frame for the short-term analysis and the entire recording duration, 24 h, for the long-term. In both cases, the outliers are filtered and the results are expressed in milliseconds. The scripts are shown in Figures 3 and 4. Note that, for the short-term analysis, only the 5 min of recording ranging from tenth to fifteenth minute of acquisition are selected, in order to remove the possible noise due to the first seconds/minutes of recording.

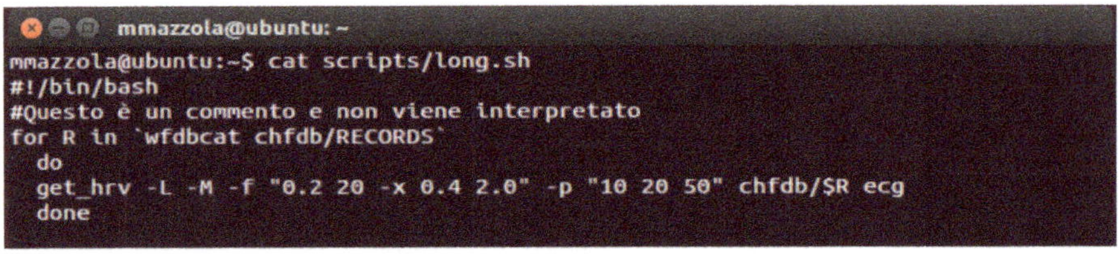

```
mmazzola@ubuntu: ~
mmazzola@ubuntu:~$ cat scripts/long.sh
#!/bin/bash
#Questo è un commento e non viene interpretato
for R in `wfdbcat chfdb/RECORDS`
  do
  get_hrv -L -M -f "0.2 20 -x 0.4 2.0" -p "10 20 50" chfdb/$R ecg
  done
```

**Figure 3.** Script for Long-Term Analysis.

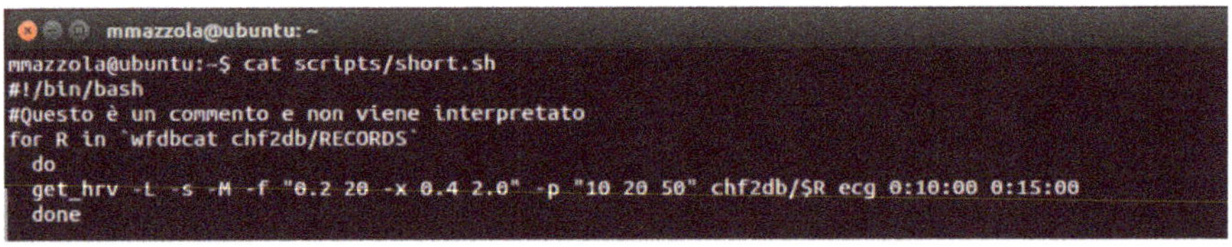

```
mmazzola@ubuntu: ~
mmazzola@ubuntu:~$ cat scripts/short.sh
#!/bin/bash
#Questo è un commento e non viene interpretato
for R in `wfdbcat chf2db/RECORDS`
  do
  get_hrv -L -s -M -f "0.2 20 -x 0.4 2.0" -p "10 20 50" chf2db/$R ecg 0:10:00 0:15:00
  done
```

**Figure 4.** Script for Short-Term Analysis.

### 3.3.3. Database Setup for Watson Analytics Analysis.

Watson Analytics (WA) is a cloud system based on regressive techniques and supervised training. The analysis dataset has been structured in a format suitable to be analyzed, as shown in Figure 5.

Each data column corresponds to an HRV parameter while each row is assigned to a different patient. Note that the last column at the right, "HF_State", represents the target prediction, which is the presence (1) or absence (0) of HF in the corresponding patient.

| NN/RR | AVNN | SDNN | SDANN | SDNNIDX | Rmssd | Pnn10 | pnn20 | pnn50 | tot_pwr | ulf_pwr | vlf_pwr | lf_pwr | hf_pwr | lf/hf | HF_State |
|---|---|---|---|---|---|---|---|---|---|---|---|---|---|---|---|
| 0.98231 | 953.09000 | 83.58880 | 76.23910 | 30.25550 | 25.27410 | 52.58590 | 18.12970 | 2.13610 | 7646.19000 | 6695.77000 | 529.78300 | 185.63000 | 235.00900 | 0.78989 | 1.00000 |
| 0.58828 | 595.97300 | 30.17450 | 81.02050 | 20.87020 | 17.03370 | 21.61640 | 4.71916 | 1.51883 | 1267.97000 | 565.83100 | 373.15800 | 144.06300 | 184.91800 | 0.77906 | 1.00000 |
| 0.94290 | 892.20600 | 54.71200 | 46.67730 | 24.11240 | 17.76940 | 33.07750 | 8.88698 | 2.52389 | 3388.65000 | 2705.19000 | 390.25000 | 124.15600 | 169.05700 | 0.73440 | 1.00000 |
| 0.97176 | 640.45300 | 52.07580 | 46.71240 | 20.52900 | 30.32520 | 33.42960 | 11.89090 | 6.35043 | 2592.57000 | 2313.55000 | 177.02400 | 37.16790 | 64.82910 | 0.57332 | 1.00000 |
| 0.98841 | 597.41500 | 50.34920 | 47.80000 | 15.51950 | 10.60370 | 31.16780 | 2.49786 | 0.19904 | 2694.50000 | 2485.25000 | 152.57900 | 28.79080 | 27.88160 | 1.03261 | 1.00000 |
| 0.65277 | 601.28000 | 64.86200 | 48.84730 | 39.06790 | 59.19110 | 52.89910 | 32.63470 | 23.18200 | 4971.86000 | 3212.92000 | 388.88700 | 387.20000 | 982.84700 | 0.39396 | 1.00000 |
| 0.95551 | 778.23500 | 57.86610 | 55.66870 | 16.12140 | 13.89710 | 43.74240 | 9.44042 | 0.31441 | 3656.22000 | 3332.15000 | 144.71000 | 76.17340 | 103.18200 | 0.73824 | 1.00000 |
| 0.97472 | 793.68700 | 59.17580 | 53.72870 | 25.03350 | 12.53730 | 34.23230 | 5.81962 | 0.31696 | 3598.09000 | 2925.94000 | 446.48400 | 132.47800 | 93.19000 | 1.42159 | 1.00000 |
| 0.98357 | 619.05200 | 33.75620 | 32.10080 | 10.38350 | 12.21500 | 27.45860 | 4.22988 | 0.59689 | 1088.53000 | 994.63600 | 40.94860 | 18.41140 | 34.52950 | 0.53321 | 1.00000 |
| 0.98927 | 484.80500 | 19.84650 | 19.56010 | 7.23082 | 7.58781 | 15.76130 | 0.71023 | 0.05299 | 374.23600 | 331.85800 | 18.56960 | 9.74586 | 14.06270 | 0.69303 | 1.00000 |
| 0.99106 | 622.80000 | 84.14780 | 83.26120 | 24.90470 | 12.86720 | 33.84800 | 8.06375 | 0.38203 | 7160.32000 | 6647.73000 | 363.17300 | 75.75240 | 83.66870 | 0.90539 | 1.00000 |
| 0.99829 | 619.80500 | 100.22600 | 96.44630 | 32.21740 | 15.02520 | 30.52880 | 9.45363 | 1.44757 | 10368.40000 | 9534.98000 | 642.71100 | 137.85900 | 53.90350 | 2.55751 | 1.00000 |
| 0.99273 | 622.47000 | 26.50500 | 25.31630 | 8.38900 | 8.67720 | 22.73980 | 0.96196 | 0.06734 | 693.16700 | 652.70000 | 22.40230 | 7.26254 | 10.80210 | 0.67233 | 1.00000 |
| 0.99645 | 768.75600 | 70.09050 | 70.31500 | 15.95480 | 15.40980 | 39.30280 | 9.89170 | 1.38673 | 4836.64000 | 4677.41000 | 81.18620 | 21.80570 | 56.24630 | 0.38768 | 1.00000 |

**Figure 5.** Abstract of current dataset in suitable format for analysis.

### 3.3.4. Data Analysis with Watson Analytics

When HRV features dataset is ready, you can start the analysis with WA. Until now the actions are performed locally; from this point on, the dataset is ready to be uploaded to the cloud and analytics operations will be performed as AaaS. Watson Analytics accepts the most common matrix formats, such as .CSV, .XSL, .XSLX. After loading the dataset, the system assigns an index value of data quality, by considering the completeness of the fields, possible presence of constant values, low number of records compared to the columns and other qualitative factors.

Now it is possible to process the dataset using the modalities offered by WA: "Assemble", "Explore", and "Predict" (see Section 2.10 above).

In Figure 6 is shown an example of use of the Assemble feature, where the distribution of Target value (HF_State) is compared to an HRV parameter (pNN20).

Valori pNN20 in relazione ad HF_state:

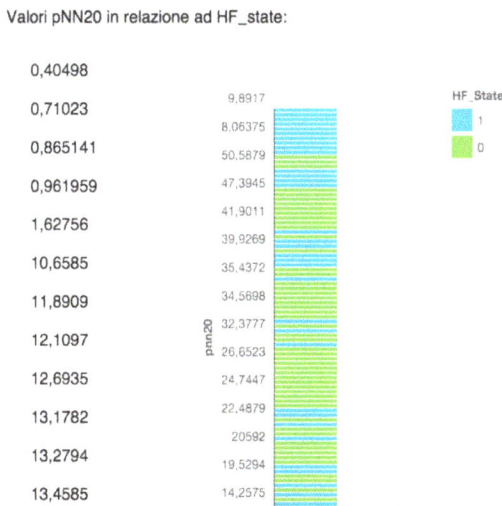

**Figure 6.** Graphic representation of the distribution of a target, based on a parameter in Assemble mode.

We can consider the Explore mode as a facilitator for the Predict mode. As seen in Figure 7, the Explore mode proposes some questions to the user in natural language. These questions are generated by relationships that WA automatically extracted from the dataset parameters (without setting any parameters as a target).

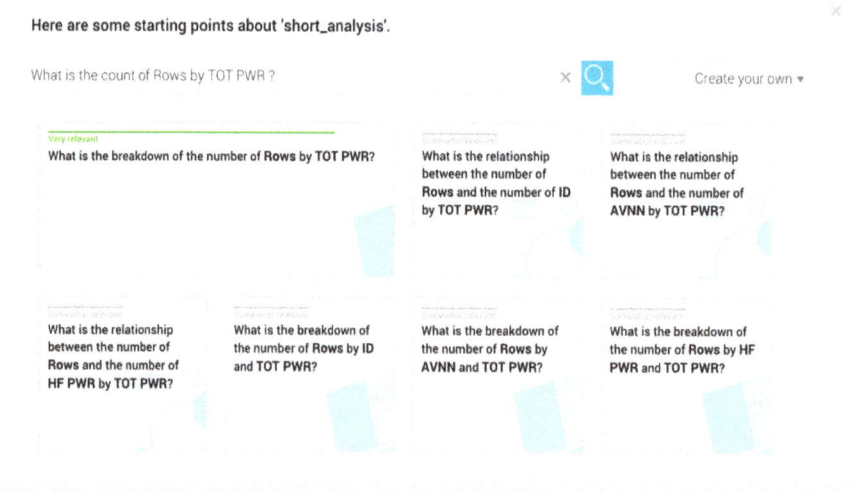

**Figure 7.** Proposals for links automatically detected in "Explore" mode.

In Explore mode, the user can also ask questions in natural language, as shown in Figure 8.

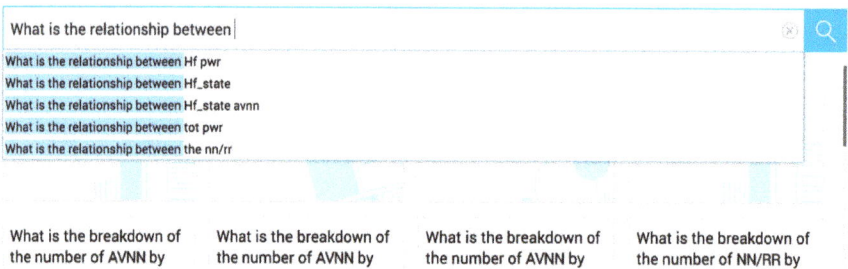

**Figure 8.** Text questions typed by user to inspect data relations or distributions.

The most interesting mode is "Predict" that allows supervised analyses by setting a prediction target. In this mode it is possible to inspect the predictive power of any other parameter.

We created two different instances of the Predict module, one for the Long-Term HRV dataset parameters and one for Short-Term HRV dataset, as explained in Section 3.3.2.

An interesting feature offered by WA is that, regardless of the type of dataset as target, it automatically chooses the most appropriate model to treat that type of data. In our case study, being HF_State a dichotomous variable, the system automatically selected the logistic regression model, as shown in Figure 9.

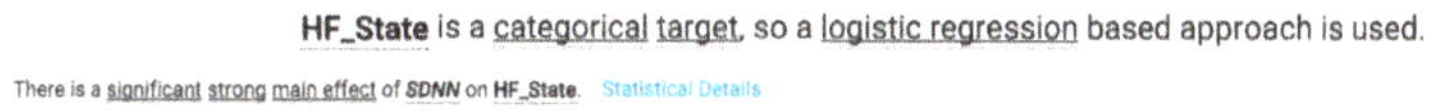

HF_State is a categorical target, so a logistic regression based approach is used.

There is a significant strong main effect of *SDNN* on **HF_State**. Statistical Details

**Figure 9.** WA has automatically selected logistic regression as the best model to deal with our data.

*3.4. Results*

In this section are shown results about the above-described AaaS use case: search for the presence of heart failure, starting from the analysis of ECG signals, using IBM Watson Analytics.

The system sets out the results both as graphics and text, in three ways:

- "single predictor": shows the predictive value of the most influent parameter
- "double predictor": the first two most predictive parameters are shown
- "combination": the various parameters are combined for a more accurate prediction.

Switching from "single predictor" to two or more predictors, the overall prediction accuracy can increase, but at the expense of the results intelligibility. In some fields of application this can be less acceptable than losing some percentage points in accuracy. Figure 10 (left box) shows these concepts.

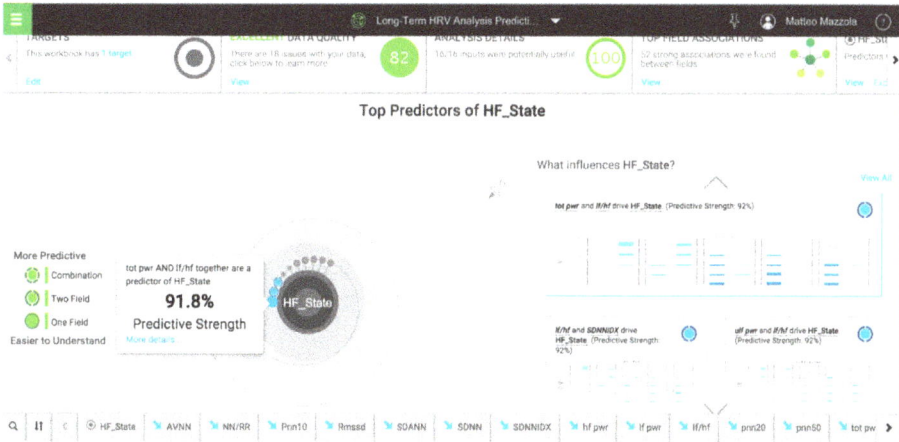

**Figure 10.** Screenshot from WA. On the left, the choice of the number of parameters to be used for the prediction, to balance intelligibility and prediction power.

3.4.1. Long-Term HRV Results

For the Long-Term HRV analysis many parameters have been spotted, having a Predictive Strength (PS) of 90% on the Target HF State, in "single predictor" mode. The most influent predictors are:

- In the Time Domain: SDNN (PS = 90%), SDANN (PS = 90%), SDNNIDX (PS = 88%)
- In the Frequency Domain: TOT_PWR (PS = 90%), ULF_PWR (PS = 90%)

Figure 11 shows, as an example, the screenshot for the parameter TOT_PWR. It can be noted that the results are displayed as numbers, text and graphics.

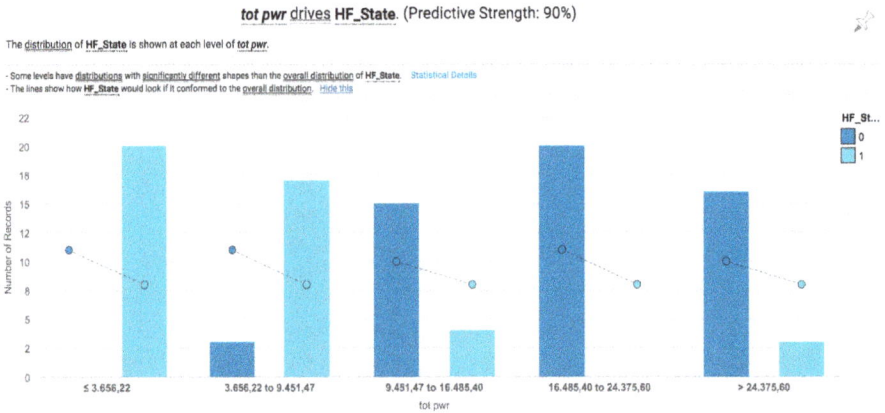

**Figure 11.** Results for Long Term Hearth Rate Variability (HRV), using TOT_PWR as single predictor.

Increasing the number of predictors to be used for the analysis, we can find many combinations with a maximum overall PS of 92%. Figure 12 shows, as an example, the combination of TOT_PWR and LF/HF.

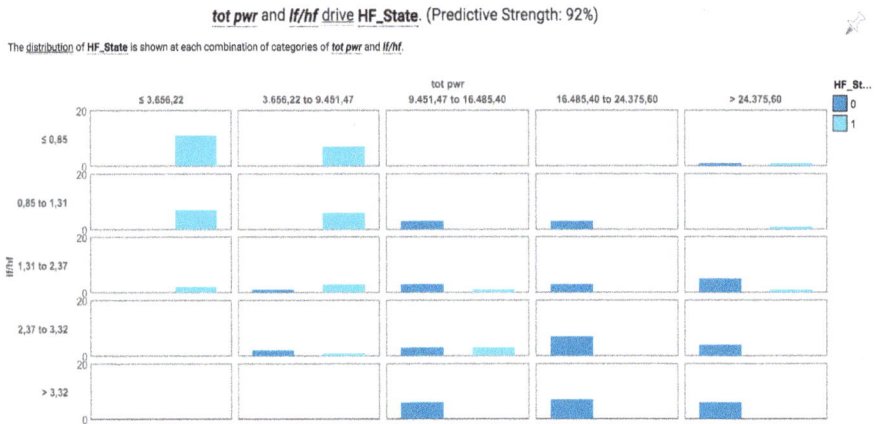

**Figure 12.** Results for Long Term HRV, using the pair TOT_PWR, LF/HF as multiple predictors.

### 3.4.2. Short-Term HRV Results

The results for the Short-Term analysis show a lower predictive power (single predictor) if compared to the Long-Term analysis. The most influent parameters on the HF_State target are:

- LF_PWR (PS = 84%)
- LF/HF (PS = 83%)
- TOT_PWR (PS = 80%)

Figure 13 shows the results for the LF_PWR parameter.

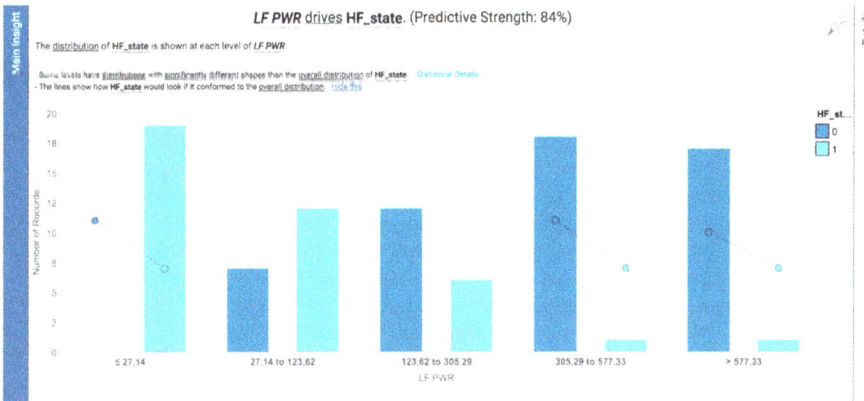

**Figure 13.** Results for Short Term HRV, using LF_PWR as single predictor.

The results are greatly enhanced by combining more predictors, achieving values similar to the Long-Term analysis:

- LF/HF combined with SDNN: PS = 94%
- LF_PWR combined with LF/HF: PS = 92%
- pNN20 combined with LF/HF: PS = 92%

Figure 14 shows the results using the combination of LF/HF and SDNN.

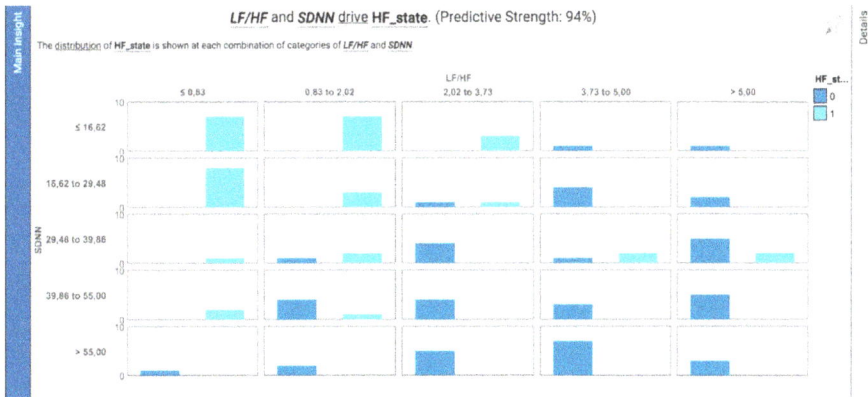

**Figure 14.** Results for Short Term HRV, using LF/HF + SDNN as multiple predictors.

## 4. Discussion on Results

The results show that the Long-Term and Short-Term HRV analyses are comparable in terms of predictive power on the detected parameters, when the target is identifying if patients are healthy or diseased (Heart Failure). The Short-Term HRV method is highly preferable, since it is much less invasive for the patient (five minutes for ECG acquisition, compared to a 24 h Holter ECG acquisition). It is also very suitable for tele-monitoring scenarios, such as those described in [29].

These results are comparable with the literature. In [23] similar results are obtained—using a static (non cloud) Classification And Regression Tree (CART) approach on MatLab—in terms of overall accuracy (>90%) and most predictive parameters (SDNN, SDANN and TOT_PWR). In [24] is described a Short-Term approach; the obtained results are similar to ours, both in terms of overall accuracy and of most effective predictors (LF/HF).

We can therefore assert that the results obtained using a cloud approach on IBM WA are comparable to the results obtained on ad hoc custom desktop platforms. The results are shown in a clear and friendly way, easily understandable also by non experts.

The main advantage of the proposed approach for the researcher is the possibility of being quickly operative, focusing only on the experiment, without taking care of hardware requirements (high computational power is needed for these analyses) or machine learning algorithms development.

From a medical point of view, the results of this study can be interpreted as the possibility to perform a preliminary and early diagnosis of HF, basing solely on the analysis of the ECG signal (accepting a certain level of uncertainty, as shown by the accuracy values).

These findings are not meant to replace the diagnostic procedures for an exhaustive diagnosis, explained in the ESC guidelines, but can be very helpful in many scenarios such as home telemonitoring for the daily monitoring of patient status.

As shown, even short term analysis has a strong predictive power: this means that the patient will benefit of the proposed approach, having to stay connected to an electrocardiograph for only 5 min (instead of 24 h).

It is very important to note that HRV analysis is based only on the progress of heart rate without any further analysis of the ECG wave form. This means that for the proposed system it is only needed a device for high quality detection of the heartbeat (for example, a 2-lead ECG measured from hands) instead of a costly and less practical 12-lead electrocardiograph. This aspect is particularly important for enabling mobile applications.

## 5. Conclusions

In this paper, after a brief introduction of the main AaaS cloud systems, we reported the experience of using a cloud-based analytics software applied to the following case study: identifying the presence of HF by analyzing the ECG signal only.

We verified that the results obtained are comparable to those found in the literature, where the same issue is addressed through custom machine learning systems, purposely developed and set up for the target case. Hence the AaaS cloud systems could be a valid alternative to local hardware and software systems for analyzing data. A major obstacle to AaaS could be transferring big datasets onto the cloud. Typical machine learning projects require the analysis of large images that can easily reach the size of 2 TB, not simply transferable onto the cloud. The model used in our case study can solve this problem by locally performing the data extraction, in order to reduce the dataset size to be transferred to the cloud.

In this study the HRV analysis has been locally performed starting from the raw ECG signal (medium size). The analysis gives back a small size vector of numeric parameters that can quickly and easily be transferred onto the cloud. This model allows you to take advantage of the full power of the AaaS approach, no matter how big is the size of the initial dataset. From a medical point of view, performing HF detection by analyzing the ECG signal only, opens the possibility of easy tele-monitoring applications (we only analyze the heart rate, not the ECG waveform, so a very basic electrocardiograph is necessary) for an early and preliminary diagnosis. Furthermore, by combining the HRV analysis with systems for assisted drugs delivering [30], it is possible to enable scenarios in which the patient is technologically aided both in diagnosis and therapy, making him more autonomous in preserving the state of his health. A more comprehensive diagnosis can then be made by performing clinical tests and following protocols as described in the ESC guidelines, and the HRV-based home tele-monitoring can be used as a daily check of patient status. Also mobile applications can highly benefit of this approach, given that it requires simple electro-medical hardware and low computational power on the local device.

**Acknowledgments:** We would like to thank IBM Italy for supporting us in the use of Watson Analytics, as well as for providing us with free student-accounts for research purposes. We would like to thank also Massimo Milli, MD, for his crucial clinical support.

**Author Contributions:** Gabriele Guidi and Ernesto Iadanza equally contributed to this manuscript. Matteo Mazzola is a student in engineering that performed the analyses using IBM Watson Analytics, supervised by Gabriele Guidi. Roberto Miniati contributed in finding the state of the art.

**Conflicts of Interest:** The authors declare no conflict of interest.

## References

1. Mell, P.; Grance, T.; Grance, T. The NIST Definition of Cloud Computing Recommendations of the National Institute of Standards and Technology. Available online: http://nvlpubs.nist.gov/nistpubs/Legacy/SP/nistspecialpublication800-145.pdf (accessed on 30 June 2016).
2. Sun, X.; Gao, B.; Fan, L.; An, W. A Cost-Effective Approach to Delivering Analytics as a Service. In Proceedings of the 2012 IEEE 19th International Conference on Web Services, Honolulu, HI, USA, 24–29 June 2012; pp. 512–519.
3. Barga, R.S.; Ekanayake, J.; Lu, W. Project Daytona: Data Analytics as a Cloud Service. In Proceedings of the 2012 IEEE 28th International Conference on Data Engineering, Arlington, VA, USA, 1–5 April 2012; pp. 1317–1320.
4. Talia, D. Clouds for Scalable Big Data Analytics. *Computer* **2013**, *46*, 98–101. [CrossRef]
5. Demirkan, H.; Delen, D. Leveraging the capabilities of service-oriented decision support systems: Putting analytics and big data in cloud. *Decis. Support. Syst.* **2013**, *55*, 412–421. [CrossRef]
6. Chen, Q.; Zeller, H. Experience in Continuous analytics as a Service (CaaaS). In Proceedings of the 14th International Conference on Extending Database Technology, Uppsala, Sweden, 21–24 March 2011; Volume 1, pp. 509–514.
7. 10 Enterprise Predictive Analytics Platforms Compared. Available online: http://www.kdnuggets.com/2013/08/10-enterprise-predictive-analytics-platforms-compared.html (accessed on 30 June 2016).
8. Enterprise Predictive Analytics Comparisons 2014. Available online: http://www.butleranalytics.com/enterprise-predictive-analytics-comparisons-2014/ (accessed on 30 June 2016).
9. SAS Analytics Home Page. Available online: Http://www.sas.com/en_us/home.html (accessed on 30 June 2016).
10. Gordon, L. Using Classification and Regression Trees (CART) in SAS ® Enterprise Miner ™ For Applications in Public Health. In *SAS Global Forum—Data Mining and Text Analytics*, Proceedings of the SAS Global Forum 2013, SanFrancisco, CA, USA, 28 April–1 May 2013; pp. 1–8.
11. Klatsky, A.L.; Hasan, A.S.; Armstrong, M.A.; Udaltsova, N.; Morton, C. Coffee, Caffeine, and Risk of Hospitalization for Arrhythmias. *Perm. J.* **2011**, *15*, 19–25. [CrossRef] [PubMed]
12. Abousalh-Neto, N.A.; Kazgan, S. Big data exploration through visual analytics. In Proceedings of the IEEE Conference on Visual Analytics Science and Technology (VAST), Seattle, WA, USA, 14–19 October 2012; pp. 285–286.
13. IBM Watson Analytics Home Page. Available online: Http://www.ibm.com/analytics/watson-analytics/ (accessed on 30 June 2016).
14. IBM Watson Analytics Community Page. Available online: https://community.watsonanalytics.com/ (accessed on 30 June 2016).
15. Watson Analytics Use Case for HR: Retaining Valuable Employees. Available online: https://www.ibm.com/blogs/watson-analytics/watson-analytics-use-case-for-hr-retaining-valuable-employees/ (accessed on 30 June 2016).
16. Watson Analytics Use Case Independence Day Edition: Fireworks and the 4th of July. Available online: http://www.scoop.it/t/gaming-analytics/p/4046911509/2015/07/02/watson-analytics-use-case-independence-day-edition-fireworks-and-the-4th-of-july (accessed on 30 June 2016).
17. Panahiazar, M.; Taslimitehrani, V.; Pereira, N.; Pathak, J. Using EHRs and Machine Learning for Heart Failure Survival Analysis. *Stud. Health Technol. Inform.* **2015**, *216*, 40–44. [PubMed]
18. Guidi, G.; Pettenati, M.C.; Miniati, R.; Iadanza, E. Random Forest for Automatic Assessment of Heart Failure Severity in a Telemonitoring Scenario. *Conf. Proc. IEEE Eng. Med. Biol. Soc.* **2013**, *2013*, 3230–3233. [PubMed]

19. Guidi, G.; Melillo, P.; Pettenati, M.; Milli, M.; Iadanza, E. Performance Assessment of a Clinical Decision Support System for analysis of Heart Failure. *IFMBE Proc.* **2014**, *41*, 1354–1357.
20. Chui, K.T.; Tsang, K.F.; Wu, C.K.; Hung, F.H.; Chi, H.R.; Chung, H.S.; Man, K.F.; Ko, K.T. Cardiovascular diseases identification using electrocardiogram health identifier based on multiple criteria decision making. *Expert Syst. Appl.* **2015**, *42*, 5684–5695. [CrossRef]
21. Boursalie, O.; Samavi, R.; Doyle, T.E. M4CVD: Mobile Machine Learning Model for Monitoring Cardiovascular Disease. *Procedia Comput. Sci.* **2015**, *63*, 384–391. [CrossRef]
22. Guidi, G.; Pettenati, M.C.; Miniati, R.; Iadanza, E. Heart Failure analysis Dashboard for patient's remote monitoring combining multiple artificial intelligence technologies. *Conf. Proc. IEEE Eng. Med. Biol. Soc.* **2012**, *2012*, 2210–2213. [PubMed]
23. Melillo, P.; Fusco, R.; Sansone, M.; Bracale, M.; Pecchia, L. Discrimination power of long-term heart rate variability measures for chronic heart failure detection. *Med. Biol. Eng. Comput.* **2011**, *49*, 67–74. [CrossRef] [PubMed]
24. Pecchia, L.; Melillo, P.; Sansone, M.; Bracale, M. Discrimination power of short-term heart rate variability measures for CHF assessment. *IEEE Trans. Inf. Technol. Biomed.* **2011**, *15*, 40–46. [CrossRef] [PubMed]
25. McMurray, J.J.V.; Adamopoulos, S.; Anker, S.D.; Auricchio, A.; Böhm, M.; Dickstein, K.; Falk, V.; Filippatos, G.; Fonseca, C.; Gomez-Sanchez, M.A.; et al. ESC Guidelines for the diagnosis and treatment of acute and chronic heart failure 2012: The Task Force for the Diagnosis and Treatment of Acute and Chronic Heart Failure 2012 of the European Society of Cardiology. Developed in collaboration with the Heart. *Eur. Heart J.* **2012**, *33*, 1787–1847. [CrossRef] [PubMed]
26. Inglis, S.C.; Clark, R.A.; McAlister, F.M.; Ball, J.; Lewinter, C.; Cullington, D.; Stewart, S.; Cleland, J. Structured telephone support or telemonitoring programmes for patients with chronic heart failure. *Cochrane Lybrary* **2010**, *8*. [CrossRef]
27. Takeda, A.; Sjc, T.; Rs, T.; Khan, F.; Krum, H.; Underwood, M. Clinical service organisation for heart failure. *Cochrane Database Syst Rev.* **2012**. [CrossRef]
28. Goldberger, A.L.; Amaral, L.A.N.; Glass, L.; Hausdorff, J.M.; Ivanov, P.C.; Mark, R.G.; Mietus, J.E.; Moody, G.B.; Peng, C.-K.; Stanley, H.E. PhysioBank, PhysioToolkit, and PhysioNet: Components of a New Research Resource for Complex Physiologic Signals. *Circulation* **2000**, *101*, E215–E220. [CrossRef] [PubMed]
29. Guidi, G.; Pollonini, L.; Dacso, C.C.; Iadanza, E. A multi-layer monitoring system for clinical management of Congestive Heart Failure. *BMC Med. Inform. Decis. Mak.* **2015**, *15* (Suppl. S3). [CrossRef] [PubMed]
30. Iadanza, E.; Baroncelli, L.; Manetti, A.; Dori, F.; Miniati, R.; Gentili, G.B. An rFId Smart container to perform drugs administration reducing adverse drug events. *IFMBE Proc.* **2011**, *37*, 679–682.

*future internet*

MDPI

*Article*

# iNUIT: Internet of Things for Urban Innovation f

Francesco Carrino [1,*], Elena Mugellini [1], Omar Abou Khaled [1], Nabil Ouerhani [2] and Juergen Ehrensberger [3]

[1]  HumanTech Institute, University of Applied Sciences and Arts Western Switzerland, Fribourg 1705, Switzerland; elena.mugellini@hefr.ch (E.M.); omar.aboukhaled@hefr.ch (O.A.K.)
[2]  Haute Ecole Arc Ing., University of Applied Sciences and Arts Western Switzerland, Neuchâtel 2000, Switzerland; nabil.ouerhani@he-arc.ch
[3]  Institute for Information and Communication Technologies, Haute Ecole d'Ingénierie et de Gestion du Canton de Vaud, University of Applied Sciences and Arts Western Switzerland, Yverdon-les-Bains 1401, Switzerland; juergen.ehrensberger@heig-vd.ch
*  Correspondence: francesco.carrino@hefr.ch; Tel.: +41-26-429-67-45

Academic Editor: Dino Giuli
Received: 13 March 2016; Accepted: 3 May 2016; Published: 11 May 2016

**Abstract:** Internet of Things (IoT) seems a viable way to enable the Smart Cities of the future. iNUIT (Internet of Things for Urban Innovation) is a multi-year research program that aims to create an ecosystem that exploits the variety of data coming from multiple sensors and connected objects installed on the scale of a city, in order to meet specific needs in terms of development of new services (physical security, resource management, *etc.*). Among the multiple research activities within iNUIT, we present two projects: SmartCrowd and OpEc. SmartCrowd aims at monitoring the crowd's movement during large events. It focuses on real-time tracking using sensors available in smartphones and on the use of a crowd simulator to detect possible dangerous scenarios. A proof-of-concept of the application has been tested at the Paléo Festival (Switzerland) showing the feasibility of the approach. OpEc (Optimisation de l'Eclairage public) aims at using IoT to implement dynamic street light management and control with the goal of reducing street light energy consumption while guaranteeing the same level of security of traditional illumination. The system has been tested during two months in a street in St-Imier (Switzerland) without interruption, validating its stability and resulting in an overall energy saving of about 56%.

**Keywords:** Internet of Things; crowd monitoring; dynamic street light management

## 1. Introduction

According to current forecasts [1], in 2050 about 86% of the European population will live in cities. This increase is a major challenge for our society that will require the development of new infrastructures and the reorganization of the urban space of tomorrow. However, the overlap between the physical territory and the digital networks along with the development of new technologies may be used to create a smart living space, cooperative, dynamic, connected and interactive, that will have an increasing impact on the economy. For instance, such an ecosystem will allow small and medium size enterprises, which usually have limited resources and skills, to create collaborative networked organizations in order to surmount their limitations through collaboration. The enormous quantity of data that will be created by this environment can be used to develop new services for the welfare and needs of its citizens with a transparent and secure access to information.

These "Smart Cities" will have to deal with issues such as the optimization of mobility, physical security, resources and waste management and urban development. As modern cities are extremely rich and complex environments, we also turned our attention towards major events (festivals, political summits, sport events, *etc.*). They offer an ideal occasion for research as they have the same features as cities but they are easier to observe since they are limited in time and space. The organization and management of big events creates many challenges that can appear as antagonists: maximizing the pleasure of the audience while ensuring the security of the event; maximizing the number of participants while providing a smooth and quick access.

Among the many issues, two areas can be considered as critical topics to manage: physical security and mobility. Physical security will probably be the major issue of our cities of the future: no activity can be developed if the security of its inhabitants is not assured. To be successful, major events must guarantee the safety of its participants through the organization of various aspects of security such as crowd management, management of panic situation, monitoring abnormal behaviors, *etc.*

Considering mobility, nowadays mobility's needs show steady growth in cities and they must be managed in terms of infrastructure and resources. When organizing a public event, proper management of the infrastructures is essential; in fact, the transportation of people and materials needs to be as quick and fluid as possible. This requires the optimization of the event topography, the distribution of relevant information in the right place for the participants, the geolocation of vehicles, intelligent parking management, *etc.*

The issues mentioned above concerning safety and mobility raise new challenges and require the development of systems easily accessible, reliable, adaptive, and that preserve privacy. The first requirement to achieve this goal is to guarantee access to sources of rich, multimedia information that can be processed and interpreted to enable reliable decision-making in real-time. The Internet of Things (IoT) is an emerging paradigm that aims to provide reliable access to heterogeneous and distributed data and may represent a good solution for the smart cities of the future. However, this new paradigm raises a number of scientific and technological challenges related to security and data protection that must be addressed comprehensively. In addition, a number of supplementary issues must be faced, such as the integration of heterogeneous data, the processing of large amounts of data, the reliability of data and of the data processing system, and the realization of an intelligent knowledge process extraction.

In this article, we present the Internet of Things for Urban Innovation (iNUIT) program. The goal of this program is to develop, through various projects, the technological "backbone" of this new urban space to improve citizens' quality of life using a multidisciplinary approach and, in particular, developing an "ecosystem" able to integrate multiple technologies, concepts and work methodologies, such as IoT, cloud computing, ubiquitous and mobile computing, decision-support systems, *etc.*

The main idea of the program is to create an ecosystem that exploits the variety of data coming from multiple sensors and connected objects installed or that will be installed on the scale of a city, in order to meet specific needs in terms of development of new services in fields such as physical security, mobility, resource management and recreation (tourism, culture, *etc.*). The vision of the program is described in Figure 1.

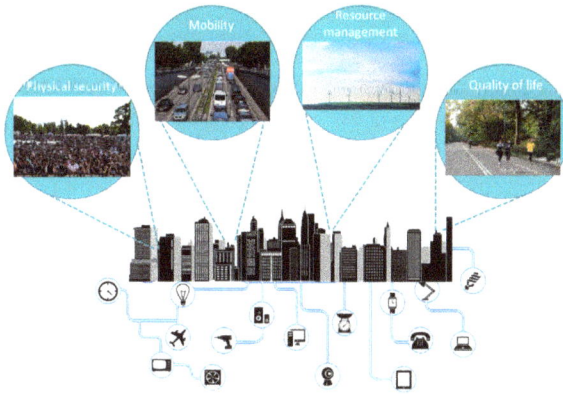

**Figure 1.** A graphic representation of the vision within the Internet of Things for Urban Innovation (iNUIT) program with main services based on a network of connected objects (Internet of Things paradigm).

In summary, the iNUIT ecosystem aims to:

- Provide citizens with services allowing the interaction with the urban environment and the interaction with their fellow citizens. Such a system could help citizens to receive the right information at the right time (e.g., to avoid traffic congestion), to provide relevant information to other citizens (e.g., in crisis situations), *etc.*
- Support the politicians through the development of methods and tools for decision-support to optimize the management of resources and waste, to improve citizen physical security, and take into account requirements of people with specific needs (e.g., older adults, disabilities), *etc.*
- Support service suppliers (such as the organizers of sport or cultural events) providing to them new ways to provide and assure their services (e.g., to better manage the flow of crowds, the traffic, and increase safety of the event).

In order to achieve these goals, iNUIT uses technologies related to IoT in specific and concrete application cases relevant for urban areas. The applications considered are distributed and involve several components (e.g., processors, sensors, actuators, computers, *etc.*) connected in networks. The iNUIT program treats the whole integration chain: acquisition, communication, processing, intelligent data analysis and presentation of results. However, scientific and technological challenges are found in all elements of this chain:

- In terms of data collection, the system should be able to support the diversity of data and the large number of "objects" to integrate: sensors, cameras, mobile phones, actuators, *etc.* The main challenge is to manage a large number of heterogeneous sensors and standardize their use to extract reliably useful data.
- The network infrastructure should consist of wireless objects that compose a self-organizing network able to convey data between the sensors and the processing infrastructure. The network infrastructure should be as generic as possible to take into account the heterogeneity of data as described in the previous point. In addition, the security of this infrastructure is critical given the confidentiality of the involved information. Finally, the infrastructure should implement complex features such as autonomous reconfiguration of the network while minimizing energy consumption.
- The massive data collection made possible by IoT raises many challenges on how to integrate, synchronize, process and recover this heterogeneous data. To do this, it is important to adopt or

develop new methods for modeling these "Big Data", new techniques of information retrieval to extract meaningful information as well as new machine learning algorithms to classify data.

Finally, given that the IoT paradigm imply collecting information from objects used in everyday life, data security and protection of privacy is of paramount importance and should be treated at all levels: from the data acquisition and the communication to the their processing that should integrate privacy preserving techniques.

Given the magnitude of the iNUIT extent, in this article, after an overview about the program's architecture, we will focus on two particular projects developed within the iNUIT vision to test the proposed architecture. The first project, called SmartCrowd, aims at monitoring the crowd's movements during large events. It focuses on the real-time tracking using GPS sensors available in smartphones and on the use of a crowd simulator to take into account topographic information to detect possible dangerous scenarios. The second project is called OpEc, which stands for "Optmisation de l'Eclairage public", the French title of the project that means "Optimization of Street Lighting". This project contributes to the resource management part of the iNUIT vision. The main objective of the project is to use IoT to implement dynamic street light management with the final goal of reducing street light energy consumption while assuring the same level of security of traditional illumination.

## 2. iNUIT Architecture

In Figure 1, we described the vision behind the iNUIT program in which different kinds of services (*e.g.*, physical security, mobility, resource management, *etc.*) relay on a bottom layer of connected objects (*e.g.*, sensors, actuators, *etc.*). A more complete vision of the iNUIT architecture is provided in Figure 2.

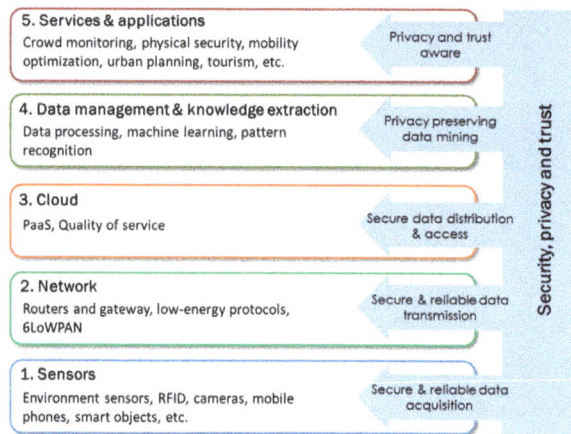

**Figure 2.** iNUIT architecture divided in five layers from the sensors (bottom) to the services (up).

The iNUIT program is structured in five layers. Security and privacy approaches are envisaged for each layer. The first layer (bottom) is the sensor layer. It contains the heterogeneous connected objects that collect and transmit data in real-time. Some of these objects may be "smarter" then others permitting the first data processing step and working as interface with the users. The chosen sensors should allow a secure and reliable data acquisition to respect security and privacy constraints (for instance, anonymizing the sent data). The network represents the second layer. It is composed of technologies and protocols that allow the transmission of the different types of data from the sensors to the cloud layer. As well as the sensors, the hardware required for the network is also distributed in the environment and therefore should have features to guarantee low electric consumption and a secure and reliable data transmission. The Cloud layer aims to make the iNUIT architecture as generic, open and flexible as possible. The "client" that will install the iNUIT system across a city (or for a particular event) should not worry about the IT infrastructure that will support the different processes (acquisition of equipment, maintenance, logistics, *etc.*). In this context, we opted to adopt a cloud solution, which supports treatment, data analysis and representation of results. Above the cloud layer, there is the layer in which are grouped the systems and algorithms that manage the data analysis (classification, knowledge extraction, *etc.*) and storage. Big data techniques should be used to take into account the large volume of data and their heterogeneity. In order to respect the users' privacy, these algorithms should be based on privacy preserving techniques (*i.e.*, techniques that do not allow the identification of a user form the available data). Finally, on the top of the pile, there is the services and applications layer that actually provides the services to the user. This layer covers also the role of interface towards the user; therefore, it should help to build the trust towards the system providing to the user the control of privacy settings (e.g., which information to share with the system and other users). Each layers of the iNUIT architecture is composed of several modules that implement the different functionalities required to provide the different services on the top layer. Figure 3 shows a more detailed view of the iNUIT architecture in which it is possible to see the modules that we implemented to develop the two projects presented in this paper: SmartCrowd and OpEc.

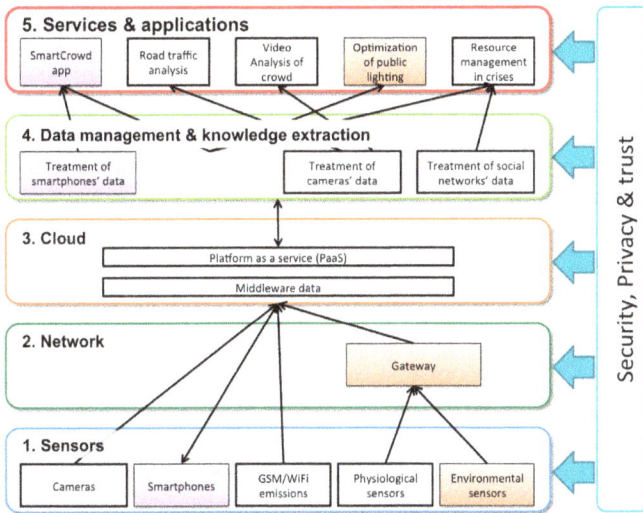

**Figure 3.** The iNUIT architecture with the modules developed within the SmartCrowd and OpEC (Optmisation de l'Eclairage public) projects tinted in purple and orange, respectively. The modules in white have been developed during other projects within the iNUIT program.

It is worth noting that the iNUIT architecture permits the implementation of very different systems to provide different services. While the Cloud layer is the core of the system flexibility, also other modules may offer functionalities that may be used by different applications. Sections 4 and 5 describe with more detail the structure of SmartCrowd and OpEc.

## 3. Related Works

What is a Smart City? The concept of Smart City is quite broad. Several definitions exist. Branchi *et al.* [2] studied several definitions and provided a more comprehensive formulation defining a Smart City as: "A space for coexistence among people who, based on the available technologies, can thrive and develop, while taking into account economic, social and environmental sustainability".

From this definition, it appears that the two essential elements that constitute a Smart City are "people" and "technology". A city cannot be detached from its citizens. They are the main actors that produce and consume all the services of the city and they are also part of the institutional factors (e.g., governance, policy, *etc.*) that should guarantee the smart growth of the community. While the citizens contribute to make the city "smart" providing these institutional and human dimensions, often the forefront role is attributed to the technology dimension and the use of "Smart Computing technologies" [3,4]. In this work, we focused more on this technological aspect.

Smart Computing Technologies are defined by Forrester [5] as "A new generation of integrated hardware, software, and network technologies that provide IT systems with real-time awareness of the real world and advanced analytics to help people make more intelligent decisions about alternatives and actions that will optimize business processes and business balance sheet results." In the Internet of Things (IoT) paradigm, many of the objects that we use daily and that surround us are connected to each other and the information and communication systems are invisibly integrated in the environment [6]. In many works, the IoT paradigm has been chosen as the ideal candidate for the implementation of the smart computing technologies needed to take into account, transparently and seamlessly, a large number of heterogeneous systems and, therefore, enabling Smart Cities. For instance, Zannella *et al.* [7] focused their analysis on Urban IoTs (*i.e.*, IoT designed to support the Smart City vision) and presented the technical solutions and best-practice guidelines used in the Padova Smart City project, in which they demonstrated its feasibility in the city of Padova, Italy. Other works have highlighted the open challenges of using IoT for Smart Cities proposing different approach or model such as hub-centric approach to the IoT [8], or a cognitive management framework for IoT (in which "cognitive" represents "the ability to dynamically select behavior through self-management, taking into account information and knowledge on the context of operation as well as policies") [9].

Similar to what was presented in the previous works, iNUIT proposes an approach based on the IoT to enable Smart Cities. iNUIT has the main goal of establishing the technological backbone for the future Smart Cities tackling concrete problems concerning its implementations: from the heterogeneous nature of the collected data to the protection of citizens' privacy. In order to do that, we attacked the problem from two different directions proposing two proof-of-concepts for two well-defined application scenarios: improving "physical security and mobility" during large events and "resource management" and, in particular, smart public illumination.

The SmartCrowd project concretizes the first direction. The goal of this project is to detect groups and dangerous situations during large events by monitoring the crowd's movement. We focused on the GPS data collected through an application running on the participants' smartphones. Since a city cannot be detached from its citizens, the idea is to change the citizens' role from passive service consumers to active and interactive actors in which they voluntarily share their information (e.g., location) in order to get new valuable services (e.g., increased safety).

The OpEc project concretizes the second direction. The goal of this project is to implement dynamic street light management using an IoT-based approach. Obviously, in this case, the citizens have a more passive role than in the previous project. Given that the goal is to manage more efficiently street lighting without reducing the safety, from the citizen point of view, the effects of the application should be as invisible as possible.

*3.1. Crowd Monitoring*

Large events may be seen as a "concentrate of urbanity". The number of participants at top world events can reach more than a million people and issues such as mobility and physical security are of paramount importance to guarantee the success of the event and to avoid serious problems. Planning the pedestrian flows and capacity plays a key role to prevent major troubles. Various techniques of estimation are used to plan pedestrian flows and capacities [10]. The required level of security can be achieved combining a precise study of the arrangement of the natural and artificial physical features of the event (*i.e.*, the topography of the event), for instance using a simulator, and real-time monitoring of the crowd [11].

Common sense suggests that the first rule should be to avoid as much as possible architectural barriers on the event site (such as obstacles, bottlenecks, hairpin turn, *etc.*), with particular attention to the emergency routes. Crowd simulators may help to detect flows in the topography of the event otherwise difficult to notice. Many types of simulators exist. Depending on the level of the detail that they take into account, simulators may be based on three different modeling approaches: microscopic, mesoscopic, and macroscopic. Macroscopic and mesoscopic approaches remove certain complexities from the equation to lower computational costs and speed up computations. On the contrary, microscopic approaches try to provide a very detailed simulation to the individuals' behavior and, even, psychological phenomena that may link each other.

An example of a macroscopic crowd simulator is represented by the use of the cellular automaton approach [12]. A cellular automaton represents the scene using a grid of cells that can be either occupied by one individual or they can be empty. Burstedde *et al.* [12] used this approach by defining a "static" floor field to direct the general movement towards an area of the grid and a "dynamic" floor field left by the pedestrians. The dynamic field was used to simulate the "herding" phenomenon [13] in which, during evacuation situations, individuals follow others individuals before them. They found that this type of simulation can recreate real pedestrian's behavior but the grid overlay may generate lockups. Due to the abstraction and restriction of movement to a grid, the computational load is low.

Example of the mesoscopic approach may be found in the works of Helbing *et al.* Helbing used "hydrodynamics" to describe the crowd's movement [14]. Then, in later work, he investigated social forces as a modeling approach for pedestrian crowd simulation and its use in normal and evacuation situations [15,16]. Essentially, considering social forces means taking into account that all individuals want to reach their destination as comfortable as possible and other social factors such as: the individuals' preference to keep a certain distance from one another (e.g., to avoid collision but also to respect the private sphere), the friends preference to move as a group, *etc.* Another interesting mesoscopic approach is described by the work of Narain *et al.* [17]. They presented a hybrid system consisting of a single-agent and fluid dynamics. In their algorithm, they overlaid a grid on the simulation scene and they used a crowd flow mechanism (called "unilateral incompressibility constraint") that calculated the flow in the grid modeling the crowd as a combination of compressible and incompressible fluid. Solving the flow for all cells of the grid provided a flow field that was used to model the velocity vector of the agents in the direction allowed by the system. This approach permitted to avoid collisions and to make the individual agents respect a minimal distance from each other. This simulation approach proved to be very performing, running a simulation with one million agents at three frames per second on an Intel Core i7 standard home computer.

Finally, using the microscopic modeling approach is possible to achieve very detailed simulation. An example is the simulator ESCAPES (Evacuation Simulation with Children, Authorities, Parents, Emotions, and Social comparison) [18]. ESCAPES was developed to simulate evacuation scenarios within buildings combining multi-agent approach (*i.e.*, individuals, families and authorities) while considering several psychological phenomena (e.g., fear, knowledge/certitude about occurred events, *etc.*). This allowed the simulation of complicated behaviors, such as: families tend to gather all family members before looking for an exit; authority agents do not exit until all the other people (families and individuals) left. They also considered that only a part of the people involved (typically, the authority agents) might be immediately aware of the danger and hold all the knowledge about the surroundings. The others may learn (and forget) facts about the environment. Finally, as knowledge, also other feelings like fear may be transmitted among people. Due to the complex nature of the model, the ESCAPES simulator allowed only a limited number of agents.

To summarize, among the several systems, a multi-agent system (microscopic approach) would probably provide the most realistic approach for crowd simulation. The more macroscopic approaches (e.g., based on fluid-dynamics paradigm) do not consider the different nuances of which a crowd of different people is composed. However, they allow the detection of potential bottlenecks or other architectural barriers and they may be usable to simulate events with a much larger number of participants with a lower computational cost. Therefore, the choice of the simulation approach should be considered depending on the aimed goal. In this work, we used the Jülich Pedestrian Simulator (JuPedSim) [19] an open framework for simulating and analyzing pedestrians' motion at a microscopic level. It allows the modeling of events (e.g., doors that open or close), the transmission of knowledge among the agents, and the measurement of the simulation in terms of density, velocity, *etc.* In terms, of adopted model, JuPedSim makes available two options: the Generalized Centrifugal Force Model (GCFM) and the Gompertz model. GCFM is a spatially continuous force-based model to simulate pedestrian dynamics, which includes an elliptical volume exclusion of pedestrians [20]. The Gompertz model is based on a continuous physical force and it may simulate social as well as physical forces in a continuous way [19]. The second approach is much more expensive in terms of processing requirements.

We extended the JuPedSim simulator to integrate a module to easily provide the topography of an event to the simulator and to calculate other parameters from the simulation able to indicate dangerous situations. In particular, we looked for the estimation of the crowd pressure, as defined by Helbing *et al.* [10]. Crowd pressure depends in local density and local speed and it takes into account the main differences among fluid dynamics and human behavior. For instance, people instead of filling all the available area tends to overcrowd some areas while leaving empty others areas.

Crowd pressure is an aspect to consider also when dealing with real-time crowd monitoring. SmartCrowd aimed to automatically detect dangerous situation (e.g., too high crowd pressure) as well as recognize groups within a crowd. Real-time crowd monitoring is usually performed using vision-based techniques. In fact, cameras provide a reliable and constant flow of video that may be analyzed by security agents or by computer-vision approaches that can assist security agents. Vision-based monitoring has several advantages such the independence (*i.e.*, it does not rely on the collaboration of the crowd) and the real-time feedback. However, it has also some important drawbacks: it depends on environmental conditions (e.g., good lighting is essential to have a clear vision of the crowd), it should consider occlusion and it suffer from poor scalability (larger spaces require a larger number of cameras and more resources to monitoring the videos [11,21]). In order to overcome these limitations, other technologies can be used to achieve real-rime crowd monitoring. With the approaching of IoT and the universalization of smartphones, new possibilities are provided by their communication properties and the use of sensors embedded in smartphones. This technology allows two different approaches to monitor crowd in real-time: in-network and on-device. In-network approaches use call data records to track the user's position by locating the cell tower who logged the phone call or message. The accuracy is about 300 m and it depends in whether the user uses the phone for activities like making a call or sending an SMS [22]. On-device approaches use the GPS sensor on the smartphone and the Wi-Fi/GSM-fingerprinting [23]. The accuracy can go up to 5 m for the GPS and around 20 m for the Wi-Fi-based technique. Along the greater precision, another advantage of the on-device method is the capability of recording data on selected, periodic basis. In addition, it permits the collection of additional information such as speed, direction and, if required, data captured by accelerometers and gyroscope. Another method that relies on the use of smartphones involves the use of Bluetooth beacons distributed in the events environment to track users positions [24]. This method provides good performance in term of precision of the localization but as well as the camera-based approach it is limited to the availability and diffusion of beacons on the area of the event.

Some studies already presented some results of crowd-monitor using smartphones-based techniques. Wirz *et al.* monitored the crowd movements during the Lord Mayors Show 2011 in London [25] using participants' smartphones. They were able to calculate the density of the crowd even if only a limited number of attendees (828 of around half a billion spectators) shared their location. Comparing their results with ground truth information obtained from video cameras used by the authorities, they were able to confirm their assumption that that walking speed is affected directly by the crowd density. A second study was made during the Züri Fäscht 2013 [26] in which around 28,000 people contributed. The authors then used the collected data to visualize information such as crowd density, crowd flow, and mobility and shared the best practices to acquire as many contributing users as possible for this kind of events.

However, the use of smartphones for real-time crowd monitoring has some challenges to face. First, it requires the active participation of the user: the installation of a specific application, the consent to share location information, *etc*. This is a veritable paradigm shift. Participants are no more mere service consumers. They become active actors that voluntarily and interactively contribute to the event, share their information with the organizers and, therefore, expect new high value-added services. Second, it suffers of some technological limitations such as batteries duration and the availability of an Internet connection. This second group of limitations can be overcome loosening the real-time constraint, for instance, reducing the frequency with which the smartphone shares the location.

Concerning the recognition of groups within a crowd many approaches exist. Being able to identify groups within the crowd may result very helpful to propose evacuation solutions according to nearest exits and to the social forces that relate people (for instance, avoiding imparting commands that may separate family and friends). This may prove to be advantageous also in terms of communications. If a given alert reach one member of a group it is probable that the information will be quickly shared. People tend to attend events in small groups rather than coming alone [27]. Ruback *et al.* [28] showed that people are likely to attend sport events in groups and they provided an overview of related works reporting that between 70% and 90% of all people present during an event came with the company of at least another person. Similar statistics are found also in other type of event or, more simply, in people walking in commercial street [29]. McPhail *et al.* [30] provided a formal definition of group in terms of proximity, direction and speed. However, it is possible to use people's location data within machine learning and clustering algorithms in order to automatically detect groups that may vary in number and size. El Mallah *et al.* [31] compared several density-based cluster algorithms in terms of accuracy and computational efficiency on a simulated database. They achieved the best result using the DBSCAN algorithm [32].

In summary, many ways to detect dangerous situations during large event exist, going from the use of simulators to the use of machine learning techniques for the real-time monitoring of the crowd. The SmartCrowd projects aims to show the possibility provided for the crowd management using an approach going in the direction of IoT. In particular, we focused our work, in the detection of dangerous situations using sensors on the smartphone, for the real-time monitoring, and a crowd simulator to detect obstacles or architectural barriers that may be present in the event location. In this article, we extend the work of El Mallah *et al.* [31] presenting the structure of iNUIT, the IoT-oriented framework that contains SmartCrowd and the first results achieved with the use of SmartCrowd during the Paleo Festival Nyon 2015.

*3.2. OpEc–Dynamic Street Light Control*

Street lighting is an essential service provided by municipalities and communities to assure the citizens and goods in public spaces. However, this essential service has non-negligible economic cost economic and also a certain ecological footprint. Regarding the economic cost of street lightening, the European cities are spending up to 60% of their electricity budget to assure this service [33]. Therefore, mastering the energy consumption of street lights is becoming an important priority for European municipalities. Leveraging Information and Communication Technologies (ICT) in managing street lights has been identified as a serious and promising approach. Dynamic street lighting (DSL) refers to allowing the adjustment of the light intensity level to a real and measured need. Different research papers on DSL have been published [34–36]. Most of the solution proposed in previous works are based on a sensor network to sense relevant ambient-related indicators like luminosity and visibility, which permits the estimation of the need for lighting intensity. Further, other previous works include presence detectors as additional sensors with the objective to consider pedestrian and road traffic activities as complementary cue for the estimation of lighting needs. Further concepts involving the citizens as main stakeholder providing feedback on lighting quality have been also proposed [37].

As for the control of light intensity, the light-emitting diode (LED) technology has allowed the conception and the implementation of numerous light control and dimming solutions. The most known technologies for dimming solutions are (1) DALI (Digital Addressable Lighting Interface), which is a protocol that permits dynamic control of luminaires; (2) "0–10 V" which is an electronic lighting control signaling system where the control signal is a Direct Current (DC) voltage that takes values between zero and ten volts. Our project takes advantage of these evolutions.

The main objective of the project is to use IoT concepts in order to implement dynamic street light management with the final goal to reduce street light energy consumption while assuring the same level of security.

## 4. SmartCrowd

### 4.1. SmartCrowd Architecture

SmartCrowd, first of all, is a proof-of-concept of iNUIT applied in a specific scenario, *i.e.*, the detection of dangerous situations during large events using data coming from the participants' smartphones and, in particular, their GPS locations. Therefore, SmartCrowd's architecture implements some modules of the iNUIT architecture, as we saw in Figure 3.

In order to make possible to monitor in real-time the crowd behavior, we developed an application for smartphone. Therefore, the smartphone represented our element in the sensors layer as well the interface towards the user in the services and applications layer. At the sensor level, the application performed the first processing step anonymizing the data, and processing them in order to transmit the information (location, speed and bearing) in the desired format. In addition, in the case of limited connection, it permitted to store the data on the smartphone to send them later when a connection was available. The data were then sent directly using a secured connection to the cloud layer. The cloud system was not developed within the SmartCrowd project but in another project related to iNUIT, therefore we do not describe here this module in detail. However, we exploited this module trough an Application Program Interface (API) that permitted to collect heterogeneous data from different smart objects. The smartphones connected to the server using this interface, they transmitted the location data and recuperated the elaborated information (such as the information to visualize in a heat map of the event). On the server side, we implemented the algorithms to calculate the crowd pressure and to recognize groups. The details of these algorithms are described in [31]. In this first prototype, the mobile app was quite simple. We developed two applications, one for iOS systems and one for Android systems. The two applications had the same structure and a very similar behavior. The application is described in the Section 4.2.

The SmartCrowd mobile app represents a service provided to the participants of the event. Obviously, the same data could be used to provide services to the organizers. In addition, also data coming from camera could be considered to enrich the results. However, we limited this first prototype to the participants' side, using only GPS as transmitted by the smartphones. Concerning information more interesting for the organizer, we developed a simulator able to take topographic information of the event and test escape situations. The SmartCrowd simulator is presented in Section 4.3.

### 4.2. SmartCrowd Mobile Application

The SmartCrowd app had the main role to implement the interface towards the users and to preprocess and send the data according to event configurations. Figure 4 shows the modules developed within the app running on the client side (regrouped by the red dashed line) and the modules running on the server side ("on the cloud") for the data analysis and management.

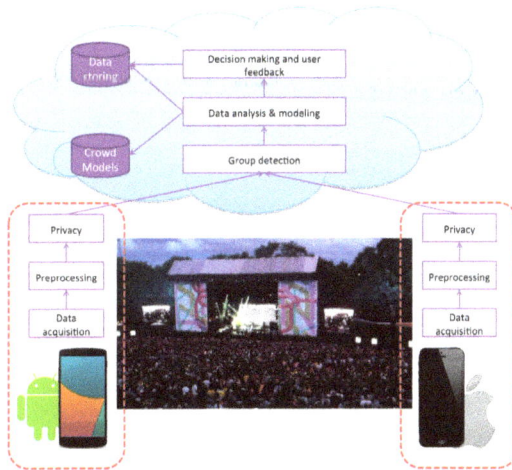

**Figure 4.** SmartCrowd app structure. The red dashed lines regroup the modules developed within the mobile app.

From the point of view of the interface, the two applications (Android and iOS) presented a very similar and simple design. Figure 5 shows three screenshots of the developed applications. Concerning the functionalities, this first prototype was quite simple. First of all, after the installation, the application showed information about the privacy management and required the consent of the user to start collecting data. To reinforce the privacy and trust aspects, the main activity of the application showed clearly the period in which the information will be collected (e.g., the duration of the event). Outside this period of time, the application stopped collecting data. The second main activity of the application offered the possibility to visualize a map of the event with the user's position, the location of the other participants that chose to share their position and the information related to the event (mainly, the location of scenes and stands position and the escape routes). Finally, a third activity showed information related to the reward. In fact, in order to foster participation in our study, we offered a reward to the participants that accepted to share (anonymously) their position information.

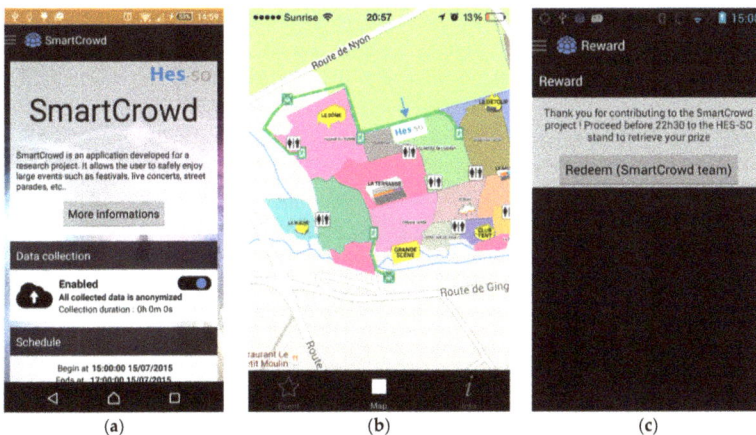

**Figure 5.** Three screenshots of the SmartCrowd application showing: (**a**) The main screen; (**b**) The map; (**c**) The reward information.

*4.3. SmartCrowd Simulator*

Our simulator extended the functionalities of the simulator JuPedSim (introduced in Section 3.1). JuPedSim is composed of three modules: JPScore, JPSvis and JPSreport. JPScore is the basic module. It calculates the trajectories considering the map configurations (walls, doors, rooms, *etc.*). JPSvis is the tool used to visualize the trajectories. Finally, JPSreport analyzes the trajectories created by the simulations and calculates the measures like density, speed, *etc.*

In order to provide to the simulator the topographical information of the event, we choose to use an online tool called Scribble Map (52 Stairs Studio Inc., Windsor, ON, Canada) [38]. This tool permitted to directly annotate in a geo-referenced map architectural elements, points of interest, gates, *etc.* simply tracking lines and polygons. Different colors corresponded to different types of elements. Moreover, it is possible to overlap an image to facilitate the annotation task. Figure 6 shows the annotation operation using Scribble Map and a map of the Paléo Festival.

**Figure 6.** A picture from the manual annotation operation.

The goal of this operation was to produce a .gpx file containing geo-referenced information about the event topography and then to translate it to the format required by the simulator with size expressed in meters. In order to do that, we developed a module able to convert the ScribbleMap's output in the right format. In particular, we used the color of the annotations to convert the physical elements to corresponding elements of the simulation; then, we converted the coordinates from the WGS84 format used in smartphones (expressed in degrees) to CH1903 (expressed in meters) and then we normalized them to avoid too large numbers. The same approach can be used to define waypoints for the agents allowing the simulation of specific paths such as escape routes (*i.e.*, routing information).

From the output point of view, JuPedSim generates a text file with the simulation results (e.g., density, velocity, *etc.*) and a Voronoi diagram to visualize these data. In addition, we added a module to visualize a heat map based on the agents' trajectories. An example of Voronoi diagrams and the heat map are shown in Figure 7.

(a)

(b)

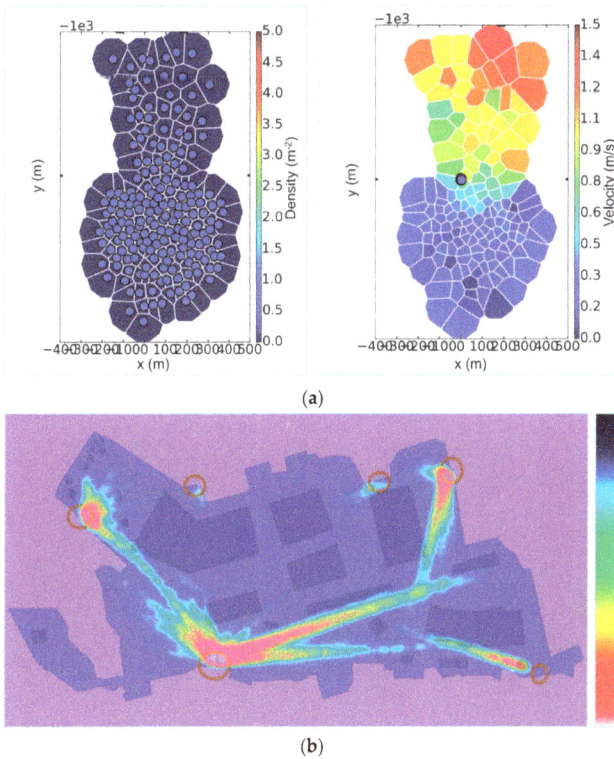

**Figure 7.** The output generated by the simulator: (**a**) two Voronoi diagrams for density and velocity of a portion of the map; and (**b**) the heat map generated by the simulation of an evacuation on the Paléo site (the orange circles indicate the exits).

The simulation also permitted seeing the displacement of the agents accordingly to one or more defined escape routes. Each person was modeled as an elliptic cylinder. Figure 8 shows two details of a running simulation.

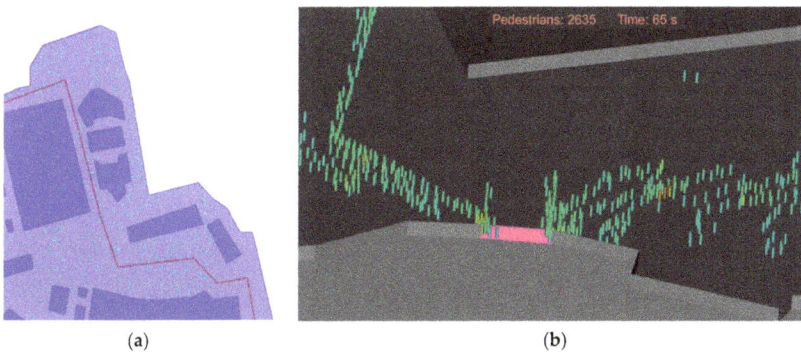

(a)

(b)

**Figure 8.** (**a**) A screenshot of the simulation scene seen form above; the red line represents the defined escaping route. (**b**) A closer look at the crowd moving toward a gate during a simulated evacuation. The color of the agent is related to the local density calculated by the simulator.

### 4.4. Paléo Festival Nyon

The Paléo Festival Nyon is an annual rock festival held in Nyon, Switzerland. In 2015 it had about 270,000 attendees distributed over seven days (*i.e.*, around 38,000 people per day). During this event, we participated in the festival by providing to volunteers the possibility of installing the SmartCrowd application.

It is worth noting that an official Paléo app existed. This application offered some static information such as the festival schedule, scenes emplacements, *etc.* and a real-time news and weather service alerts. Unfortunately, we could not integrate the SmartCrowd app within the official Paléo app. Therefore, to foster contributions to our study, we provided a reward to the participants that accepted to install the app and share their location information. In order to limit the consumption of battery, we limited the duration of the data collection to an hour. We performed our test on three days during the festival.

- First day. Period: 8:30–9:30 p.m. Apps downloaded: 18 iOS, 10 Android. 140,765 points collected.
- Second day. Period: 7:30–8:30 p.m. Apps downloaded: 21 iOS, 14 Android. 109,962 points collected.
- Third day. Period: 8:30–9:30 p.m. Apps downloaded: 25 iOS, 12 Android; 331,218 points collected.

Figure 9 shows some data and statistics about the information collected during the third day.

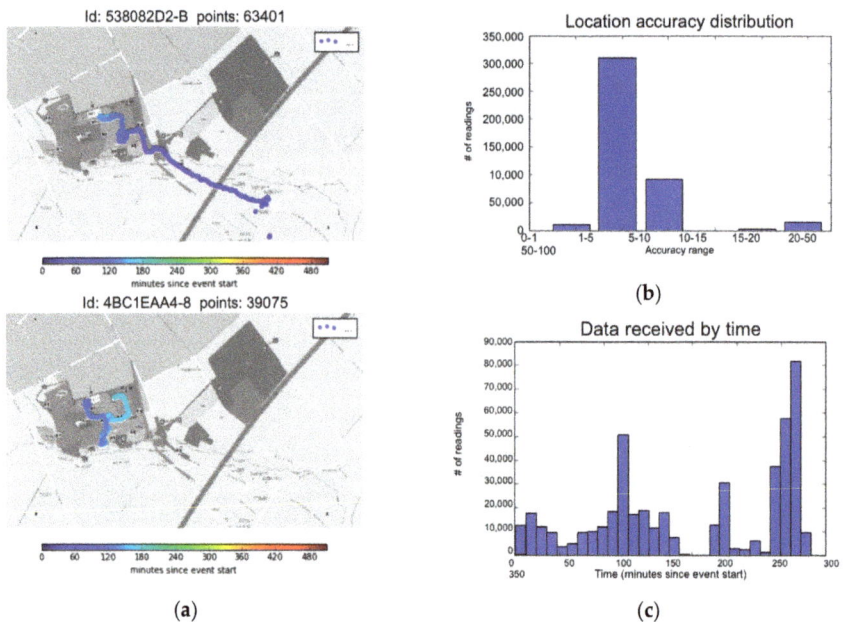

**Figure 9.** (**a**) Two users' location points during the one-hour collect time. (**b**) The accuracy distribution and (**c**) the time distribution of data collected during the event.

Even if the number of the applications download was quite limited, this study permitted us to get a first evaluation of the proposed approach. The use of smartphones may provide quite accurate data. Around 73% of the data had an accuracy estimated at 5–10 m while 22% was 10–15 m. This permitted reconstructing the displacement during the event as shown in Figure 9.

Unfortunately, the number of participants was too limited to be used to compute measures such as crowd pressure or groups detection.

To complete these data, during the event we interviewed 172 participants. Almost all the interviewed people had a smartphone even if some of them could not install our application for "technical" reasons (e.g., unavailable 3G or 4G connections, incompatible phone, lost account password, battery limitations, *etc.*). The main result from the survey was that "security" was not perceived as sufficient motivation to install the application. Most of the participants justified this answer explaining that they were "Paléo experts" (*i.e.*, with several previous participations in the festival) and, therefore, they did not need an application to help them in case of danger. Other participants did not consider the application useful for "their case", since they planned to participate in just one day of the festival.

Each user had a quite personal vision about privacy and confidentiality. Someone defined himself as almost "paranoid", claiming to not install any application that can access to his or her data. On the contrary, other participants claimed not caring at all about privacy since "we are all spied on anyway". However, most of people claimed to be sensitive to the problem but also to be open to reduce their privacy in exchange for services perceived as useful.

Regarding the Paléo official application, 48% of the interviewed participants had it installed, mainly to be informed about schedules and concert locations and alert services in case of bad weather and changes in show times. Given that the official application already requires the consent of the users to share location information for some of its features, it seems that a viable way to use smartphone' data to increase the event security could be to integrate the SmartCrowd functionalities in the event official app.

Concerning the people that did not install the official app, 83% of them justified their choice with the lack of perceived usefulness ("the same information are available on the paper pamphlet"), 10% were not aware of the existence of the application, and 7% did not install it for technical reasons. In addition, we tried to investigate which motivations could encourage them to install the application but there was not a clear answer. They expressed interest towards "context aware" features (e.g., find the nearest bar or the toilette with lower waiting time) while they almost completely discarded safety features and "find-my-friends"-like features ("we have whatsapp for that"). These results are in line with previous research, such as [26], in which they found that the main interest to install the festival application was to get the event's map and program, followed by a collection game. Similar to the results of our survey, they found that participants largely overlooked the safety-related concepts of the app. This seems to confirm that a "gamification" approach might encourage the users to share their data more than safety-oriented applications that are perceived as too "abstract" to grasp. For what concerns context aware functionalities, a deeper study is required. Participants seemed interested to the concept but only the actual use of such features could prove if they respond to actual needs.

A second phase of tests is scheduled for the 2016 festival edition in which we are working to integrate the SmartCrowd application with the official Paléo app.

## 5. OpEc

### 5.1. OpEc Architecture

The system view of the dynamic street light control and management solution is illustrated on Figure 10. The solution takes advantage of the IoT paradigm in order to collect sensor data, analyze them, make a decision on the needs of street light intensity and send commands to an actuator that control the light intensity of luminaires. The system is composed of the following parts.

- Central platform: Analyzes the data sensed by the sensors and relayed to a central platform. Its main objective is to "understand" the environmental context and, thus, to determine the real needs for lighting intensity, *i.e.* intense/low activity, good/bad visibility, *etc.*
- Sensors that deliver:
  - environmental indicators, e.g., luminosity, weather forecast; and
  - activity indicators, e.g., pedestrian, traffic;

- Luminaires: Receive the command to regulate the light intensity according to the determined needs.

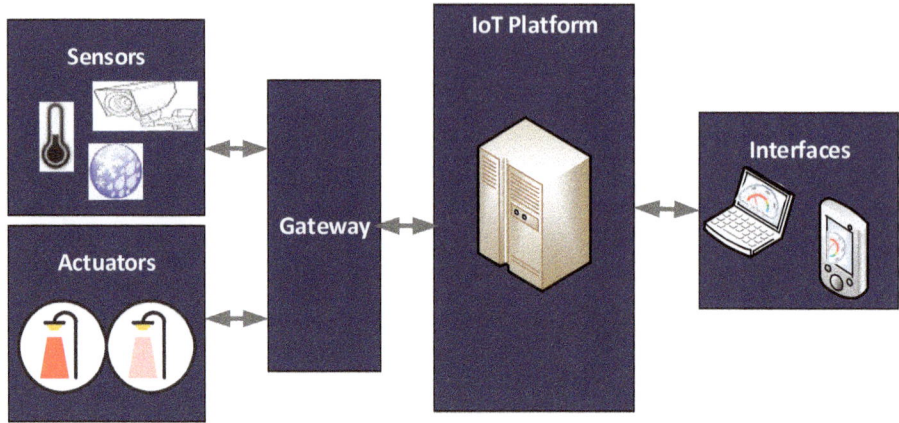

**Figure 10.** Dynamic street lighting system overview.

The different parts will be presented in detail below.

The proposed light management and control solution is based on technological components inspired from the Internet of Things (IoT) paradigm, presented in [39]. A set of sensors are deployed in the field to measure street light relevant indicators which are processed and analyzed with the objective to accurately estimate the real need related to light intensity for a given context. Finally, a command is sent to an individual or a group of luminaires for dynamically setting the needed light intensity. The IoT platform built around a set of building blocks for each layer of the IoT technological stack is used to integrate the DSL solution.

The platform is conceived and built to flexibly and seamlessly integrate new communicating objects with minimal effort. Indeed, the platform addresses the IoT fragmentation challenge, which is caused the lack of standards in communication protocols and data format to conceive and build IoT solutions. Our platform is implemented around the following modules.

Communicating Devices. The sensors and actuators are integrated on a single embedded device able to forward the measured values and to receive the commands.

Gateway. A ubiquitous processing unit capable of communicating with the end devices and to push the measured values to the central platform. Further, the gateway is able to pull the commands to the end devices. In order to facilitate the connection and the integration of different devices that use different communication protocol, the Gateway relies on communication agents that are able to understand various protocols and extract relevant sensor data from the protocol's payload. These communication agents allow the seamless integration of communicating devices. On top of the Java VM, the framework Open Services Gateway initiative (OSGi) [40] has been installed. OSGi is a java-based execution environment suitable for deploying and running modular and loosely coupled services implemented in the form of bundles. Further, OSGi allows the realization and the execution of applications, which are conform to the Service Oriented Architecture (SOA) principles and able to run on a resources-constrained hardware. In order to avoid the manual implementation and management of low-level services like device abstraction, connectivity management, *etc.*, the proposed solution makes extensive use of the Kura framework [41].

Broker. It is a message broker capable of forwarding the data between the gateway and the central platform based on the «publish/subscribe» principle. The underlying asynchronous communication mechanism assures a high performance of the data transfer.

Rules engine. The dynamic street light application relies on an engine based on Drools (Red Hat Inc. or third-party contributors, Raleigh, NC, USA). It combines the different sensor data and infers a decision regarding the light intensity to configure on the luminaire. Figure 11 illustrates the implemented sensor fusion algorithms using a decision tree.

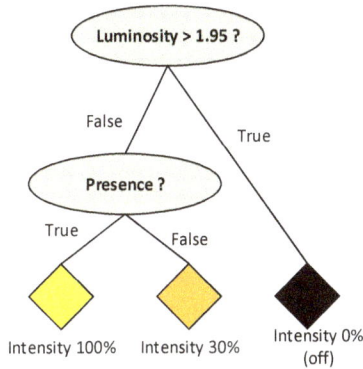

**Figure 11.** Decision tree implemented in Drools.

Zigbee-based Outdoor Light Controller (OLC). In order to dynamically and remotely control the intensity of the luminaires using the central IoT platform and the associated Gateway, we need an OLC that receives, on one side, via ZigBee protocol, the intensity percentage that has been calculated by the decision making module of the platform, and sends on the other side, via a wired DALI or 0–10 V interface, the regulation signal to the luminaire.

*5.2. OpEc Result*

After several tests carried out in a lab environment to tune the different modules of the dynamic street light control solution, we have deployed the solution at the city of St-Imier in Switzerland. We have chosen a street located near our Lab, which name is "Rue de la Cible". The deployed instance of the solution is composed of the following elements:

- IoT Platform installed on a virtual server at our lab and the Gateway installed on a Rasberry Pi (Raspberry Pi Foundation, Caldecote, United Kingdom) running Ubuntu linux.
- The ZigBee enabled Outdoor Light Controller with the DALI converter activated.
- A luminaire equipped with a DALI slave and connected to a 24/7 power supply of the street lighting electric network of the city.
- A camera connected to a Raspberry Pi that detects the activities on the street based on motion detection.
- A weather station connected to a Raspberry Pi measuring mainly the ambient luminosity.

Table 1 shows the parameters used for the configuration of the decision tree that combines the different sensor data and infers the needed light intensity.

**Table 1.** Configuration of the intensity variation rule.

| Intensity variation rules | | Luminosity | |
|---|---|---|---|
| | | Bright | Dark |
| Activity | No | 0% | 30% |
| | Yes | 0% | 100% |

The test has been run for two months during November and December 2015. The system has run without any interruption and the stability has been validated.

Further, the overall energy saving for the instrumented luminaire is around 56%. As shown in Figure 12, the intensity of the luminaire is set to 30% during 80% of the nighttime. Only 20% of the time the luminaire operates at 100% of its maximal capacity.

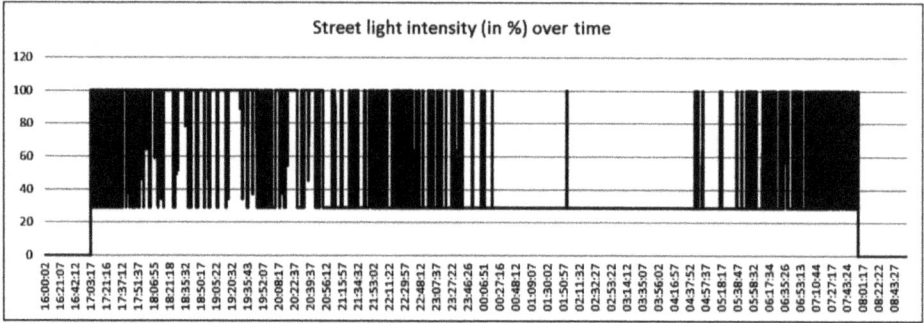

**Figure 12.** Luminaire intensity level variation during one night.

In addition to this quantitative evaluation, qualitative tests have also been run by our team in order to validate the reliability of the motion detectors and the reactivity of the light intensity to road activities (pedestrians and cars). These tests have clearly shown that the level of security perception offered by the dynamic light solution is comparable to continuous maximal illumination situations.

## 6. Conclusions

In this article, we presented the iNUIT program and its architecture. The goal of the project is to use the Internet of Thing paradigm to enable the idea of Smart City offering new services in the direction of physical security, resource management, mobility, and quality of life. To validate the idea, we presented two projects developed within the context of the iNUIT vision: SmartCrowd and OpEc.

SmartCrowd implemented a part of the iNUIT infrastructure with the goal of monitoring crowd's movement during large events. In particular, the projects used sensor embedded in users' smartphones to detected groups in the crowd and estimate parameters such as crowd pressure to detect possible dangerous situations in real-time. In addition, we extended an existing simulator to analyze offline the impact of event topography (*i.e.*, the arrangement of the natural and artificial physical features) site on security. The system was tested during the Paléo Festival Nyon 2015. The test showed that is possible to use smartphones' sensors to anonymously monitor the crowd's movement in real-time with a precision up to 5 m. However, in order to get an exhaustive vision of the crowd's behavior (such as groups recognition or a precise estimation of crowd pressure), it is necessary the contribution of a large part of the participants. According to a survey that involved 172 participants, a good solution could be represented by the integration of the services offered by SmartCrowd within the event's official application. In the future, it will be interesting to study the possibilities offered by such an application in terms of contextualized services and services offered to the event organizers and law enforcement. Videos about the simulation and the data tracked during the festival are available online [42].

Regarding the OpEc project, the current system relies mainly on environmental and traffic indicators to estimate the needs for street light intensity. It then controls the luminaires accordingly. Quantitative tests have been carried out on real world scenario using real street light luminaires deployed on one street of a Swiss city. These tests clearly showed the relevance of the solution. The overall energy saving when using our solution is measured and it amounts to 56% for the considered scenario. Qualitative tests have been also run with humans and no major impact on the security perception of the light intensity dynamic have been reported, which reinforces the potential of such systems to bring part of the solution towards low carbon cities of the future.

**Acknowledgments:** The research program iNUIT is part of the large-scale thematic research programs of the University of Applied Sciences and Arts of Western Switzerland (HES-SO). The authors gratefully acknowledge funding from HES-SO during the period of 2013–2020. In addition, the authors would like to thank Andres Perez-Uribe and Julien Rebetez for the valuable contribution in the development and data analysis activities during the SmartCrowd project.

**Author Contributions:** J.E., E.M. and N.O. participated to the conception of the iNUIT architecture; F.C., E.M. and O.A.K. managed the SmartCrowd project, performed the experiment and analyzed the data; N.U. managed the OpEc project, performed the experiment and analyzed the data; and all the authors contributed to the writing of the paper.

**Conflicts of Interest:** The authors declare no conflict of interest.

## Abbreviations

The following abbreviations are used in this manuscript:

| | |
|---|---|
| API | Application Program Interface |
| DALI | Digital Addressable Lighting Interface |
| DC | Direct Current |
| ESCAPES | Evacuation Simulation with Children, Authorities, Parents, Emotions, and Social comparison |
| GCFM | Generalized Centrifugal Force Model |
| iNUIT | Internet of Things for Urban Innovation |
| IoT | Internet of Things |
| OLC | Outdoor Light Controller |
| OpEc | Optmisation de l'Eclairage public |
| OSGi | Open Services Gateway initiative |
| SOA | Service Oriented Architecture |

## References

1. Caragliu, A.; del Bo, C.; Nijkamp, P. Smart Cities in Europe. *J. Urban Technol.* **2011**, *18*, 65–82. [CrossRef]
2. Branchi, P.; Fernández-Valdivielso, C.; Matias, I. Analysis Matrix for Smart Cities. *Future Internet* **2014**, *6*, 61–75. [CrossRef]
3. Nam, T.; Pardo, T.A. Conceptualizing smart city with dimensions of technology, people, and institutions. In Proceedings of the 12th Annual International Digital Government Research Conference on Digital Government Innovation in Challenging Times—dg.o '11, College Park, MD, USA, 12–15 June 2011; ACM Press: New York, NY, USA, 2011.
4. Chourabi, H.; Nam, T.; Walker, S.; Gil-Garcia, J.R.; Mellouli, S.; Nahon, K.; Pardo, T.A.; Scholl, H.J. Understanding Smart Cities: An Integrative Framework. In Proceedings of the 2012 45th Hawaii International Conference on System, Maui, HI, USA, 4–7 January 2012; pp. 2289–2297.
5. Washburn, D.; Sindhu, U. Helping CIOs Understand "Smart City" Initiatives. *Growth* **2009**, *17*, 1–16.
6. Gubbi, J.; Buyya, R.; Marusic, S.; Palaniswami, M. Internet of Things (IoT): A vision, architectural elements, and future directions. *Future Gener. Comput. Syst.* **2013**, *29*, 1645–1660. [CrossRef]
7. Zanella, A.; Bui, N.; Castellani, A.; Vangelista, L.; Zorzi, M. Internet of Things for Smart Cities. *IEEE Internet Things J.* **2014**, *1*, 22–32. [CrossRef]
8. Lea, R.; Blackstock, M. Smart Cities: An IoT-centric Approach. In Proceedings of the 2014 International Workshop on Web Intelligence and Smart Sensing—IWWISS '14, Saint Etienne, France, 1–2 September 2014; ACM Press: New York, NY, USA, 2014; pp. 1–2.
9. Vlacheas, P.; Giaffreda, R.; Stavroulaki, V.; Kelaidonis, D.; Foteinos, V.; Poulios, G.; Demestichas, P.; Somov, A.; Biswas, A.; Moessner, K. Enabling smart cities through a cognitive management framework for the internet of things. *IEEE Commun. Mag.* **2013**, *51*, 102–111. [CrossRef]
10. Helbing, D.; Johansson, A.; Al-Abideen, H.Z. Dynamics of crowd disasters: An empirical study. *Phys. Rev. E* **2007**, *75*. [CrossRef] [PubMed]

11. Zhan, B.; Monekosso, D.N.; Remagnino, P.; Velastin, S.A.; Xu, L.-Q. Crowd analysis: A survey. *Mach. Vis. Appl.* **2008**, *19*, 345–357. [CrossRef]
12. Schadschneider, A. Cellular Automaton Approach to Pedestrian Dynamics—Theory. *Pedestr. Evacu. Dyn.* **2001**, *11*, 75–86.
13. Stroehle, J. How do pedestrian crowds react when they are in an emergency situation–models and software Pedestrian behavior. Available online: http://guava.physics.uiuc.edu/~{}nigel/courses/569/Essays_Fall2008/files/Stroehle.pdf (accessed on 10 May 2016).
14. Helbing, D. A fluid dynamic model for the movement of pedestrians. *Complex Syst.* **1992**, *6*, 391–415.
15. Helbing, D.; Molnár, P. Social Force Model for Pedestrian Dynamics. *Phys. Rev. E* **1995**, *51*. [CrossRef]
16. Helbing, D.; Farkas, I. Simulation of pedestrian crowds in normal and evacuation situations. *Pedestr. Evacu. Dyn.* **2002**, *21*, 21–58.
17. Narain, R.; Golas, A.; Curtis, S.; Lin, M.C. Aggregate dynamics for dense crowd simulation. *ACM Trans. Graph.* **2009**, *28*. [CrossRef]
18. Tsai, J.; Fridman, N.; Bowring, E.; Brown, M.; Epstein, S.; Kaminka, G.; Marsella, S.; Ogden, A.; Rika, I.; Sheel, A.; *et al.* ESCAPES: Evacuation simulation with children, authorities, parents, emotions, and social comparison. In Proceedings of the 10th International Conference Auton, Taipei, Taiwan, 2–6 May 2011; Volume 2, pp. 457–464.
19. Ulrich, A.U.K.; Chraibi, M.; Zhang, J.; Lammel, G. JuPedSim: An open framework for simulating and analyzing the dynamics of pedestrians. In Proceedings of the 3rd Conference of Transportation Research Group of India, Kolkata, India, 17–20 December 2015.
20. Chraibi, M.; Seyfried, A.; Schadschneider, A. Generalized centrifugal-force model for pedestrian dynamics. *Phys. Rev. E* **2010**, *82*. [CrossRef] [PubMed]
21. Gong, S.; Loy, C.C.; Xiang, T. Security and Surveillance. In *Visual Analysis of Humans*; Moeslund, T.B., Hilton, A., Krüger, V., Sigal, L., Eds.; Springer London: London, UK, 2011; pp. 455–472.
22. Becker, R.A.; Caceres, R.; Hanson, K.; Loh, J.M.; Urbanek, S.; Varshavsky, A.; Volinsky, C. A Tale of One City: Using Cellular Network Data for Urban Planning. *IEEE Pervas. Comput.* **2011**, *10*, 18–26. [CrossRef]
23. Kim, D.H.; Kim, Y.; Estrin, D.; Srivastava, M.B. SensLoc. In Proceedings of the 8th ACM Conference on Embedded Networked Sensor Systems—SenSys '10, Zurich, Switzerland, 3–5 November 2010; ACM Press: New York, NY, USA, 2010; p. 43.
24. Versichele, M.; Neutens, T.; Delafontaine, M.; van de Weghe, N. The use of Bluetooth for analysing spatiotemporal dynamics of human movement at mass events: A case study of the Ghent Festivities. *Appl. Geogr.* **2012**, *32*, 208–220. [CrossRef]
25. Wirz, M.; Franke, T.; Roggen, D.; Mitleton-Kelly, E.; Lukowicz, P.; Tröster, G. Probing crowd density through smartphones in city-scale mass gatherings. *EPJ Data Sci.* **2013**, *2*. [CrossRef]
26. Blanke, U.; Troster, G.; Franke, T.; Lukowicz, P. Capturing crowd dynamics at large scale events using participatory GPS-localization. In Proceedings of the 2014 IEEE Ninth International Conference on Intelligent Sensors, Sensor Networks and Information Processing (ISSNIP), Singapore, 21–24 April 2014; pp. 1–7.
27. Johnson, N.R. Panic at "The Who Concert Stampede": An Empirical Assessment. *Soc. Probl.* **1987**, *34*, 362–373. [CrossRef]
28. Ruback, R.B.; Collins, R.T.; Koon-Magnin, S.; Ge, W.; Bonkiewicz, L.; Lutz, C.E. People Transitioning Across Places: A Multimethod Investigation of How People Go to Football Games. *Environ. Behav.* **2013**, *45*, 239–266. [CrossRef]
29. Moussaïd, M.; Perozo, N.; Garnier, S.; Helbing, D.; Theraulaz, G. The Walking Behaviour of Pedestrian Social Groups and Its Impact on Crowd Dynamics. *PLoS ONE* **2010**, *5*, e10047. [CrossRef] [PubMed]
30. McPhail, C.; Wohlstein, R.T. Using Film to Analyze Pedestrian Behavior. *Sociol. Methods Res.* **1982**, *10*, 347–375. [CrossRef]
31. Mallah, J.E.; Carrino, F.; Abou Khaled, O.; Mugellini, E. Crowd Monitoring—Critical situations prevention using smartphones. In *Distributed, Ambient, and Pervasive Interactions*; Springer International Publishing: Cham, Switzerland, 2015; Volume 9189, pp. 496–505.
32. Ester, M.; Kriegel, H.-P.; Sander, J.; Xiaowei, X. A density-based algorithm for discovering clusters in large spatial databases with noise. *InKdd* **1996**, *96*, 226–231.

33. European Commission. Lighting the Cities: Accelerating the Deployment of Innovative Lighting in European Cities. Available online: https://ec.europa.eu/digital-single-market/en/news/new-commission-report-lighting-cities-accelerating deployment-innovative-lighting-european (accessed on 10 May 2016).
34. Husin, R.; Al Junid, S.A.M.; Majid, Z.A. Automatic Street Lighting System for Energy Efficiency based on Low Cost Microcontroller. *Int. J. Simul. Syst. Sci. Technol.* **2012**, *13*, 29–34.
35. Sung, W.-T.; Lin, J.-S. Design and Implementation of a Smart LED Lighting System Using a Self Adaptive Weighted Data Fusion Algorithm. *Sensors* **2013**, *13*, 16915–16939. [CrossRef]
36. Kapgate, D. Wireless Streetlight Control System. *Int. J. Comput. Appl.* **2012**, *41*, 1–7. [CrossRef]
37. Ouerhani, N.; Pazos, N.; Aeberli, M.; Senn, J.; Gobron, S. Dynamic Street Light Management–Towards a citizen centered approach. In Proceedings of the 3rd International Conference on Hybrid City, Athens, Greece, 17–19 September 2015.
38. 52 Stairs Studio Inc. Scribble Map. Available online: http://www.scribblemaps.com/ (accessed on 2 March 2016).
39. Final architectural reference model for the IoT, version 3.0. Available online: http://www.iot-a.eu/public/public-documents/d1.5/at_download/file (accessed on 6 May 2016).
40. OSGi Alliance. Available online: http://www.osgi.org (accessed on 7 March 2016).
41. Kura. Available online: https://eclipse.org/kura/ (accessed on 7 March 2016).
42. SmartCrowd video playlist. Available online: https://goo.gl/gvVX5x (accessed on 6 May 2016).

*future internet*

MDPI

*Article*

# A Point of View on New Education for Smart Citizenship

f

## Cristina Martelli

Department of Statistics, Informatics and Applications, University of Florence, 50134 Firenze, Italy;
cristina.martelli@unifi.it; Tel.: +39-329-360-3327

Academic Editor: Dino Giuli
Received: 21 September 2016; Accepted: 17 January 2017; Published: 1 February 2017

**Abstract:** Smart cities and intelligent communities have an ever-growing demand for specialized smart services, applications, and research-driven innovation. Knowledge of users' profiles, behavior, and preferences are a potentially dangerous side effect of smart services. Citizens are usually not aware of the knowledge bases generated by the IT services they use: this dimension of the contemporary and digital era sheds new light on the elements concerning the concept of citizenship itself, as it affects dimensions like freedom and privacy. This paper addresses this issue from an education system perspective, and advances a non-technical methodology for being aware and recognizing knowledge bases generated by user-service interaction. Starting from narratives, developed in natural language by unskilled smart service users about their experience, the proposed method advances an original methodology, which is identified in the conceptual models derived from these narratives, a bridge towards a deeper understanding of the informative implications of their behavior. The proposal; which is iterative and scalable; has been tested on the field and some examples of lesson contents are presented and discussed.

**Keywords:** smart education; social innovation; statistical information systems; data reuse; transparency

---

## 1. Introduction

New technologies have a huge impact on social life [1,2], changing (i) the quality of personal relationships [3,4]; (ii) the concept of proximity [5]; (iii) the idea of authority; (iv) privacy; (v)liberty; (vi) economy [6]; and (vii) democracy.

Many authors have reflected on the impact of new technologies on personal lives and on the concept of being a citizen: as new technologies are always information-rooted, considerations have been made in terms of the right of being skilled enough to cope with modernity [7,8] and on the respect of rights in the digital society.

The role of smart technologies in social participation [9–11] has also been widely explored and discussed, as well as how they improve and empower citizens' initiative, proactivity, and involvement efforts.

The aim of this paper is to address the issue of "being an active and conscious citizen" (in all the accepted meanings of this expression) from a different perspective. Information technologies use and create data spaces: a plurality of repositories and databases keep citizens' memories, creating a sort of *digital territory* that any modern citizen is obliged to inhabit. Among traditional citizenship dimensions, new rights, linked to these new digital territories, have to be granted: awareness of potential dangers is the main element to achieve, as citizens usually fail to recognize the smart service costs, mainly focused on privacy and freedom.

The underlying idea is that citizens cannot be fully integrated into a modern society without a clear image of their position and role in digital spaces. In this sense, when asking or using smart services and intelligent territories, citizens ought to be aware (i) of the processes (mainly informative and induced by the services they use) involving communities and individuals; and (ii) of their real body boundaries in this unprecedented information territory. The effects of the presence of a person in digital territories may, in fact, deeply impact their lives, relationships and personal security.

The deep gap existing between stakeholders' languages may hamper these goals. Technicians in charge of the design and construction of smart infrastructure and services are (and often are pleased to be) beyond the comprehension of other citizens: a consequence is that an increasing component of common experiences is outside the citizens' control, vigilance, and awareness. In this work, we advance the idea that natural language, when linked to technology basics, may be a powerful instrument for achieving social consciousness and broad recognition of technology costs and benefits. This point is hardly surprising as linguistic competency is traditionally a core capability for full citizenship [12,13]: in this paper, this idea is developed and updated for the smart citizens' benefit.

In this perspective, an original educational proposal is advanced and some experiences discussed. The aim is to achieve a methodology, based on conceptual modeling, for addressing new citizenship issues, adopting different and converging perspectives: narrations, data, IT, and measures. The final goal is to induce in young students (and in their teachers) the idea that, in order to be an active citizen in modern society, all of these skills must act together, sharing a common language [14].

The paper is organized as follows: after a brief focus on the complexity paradigm, the concept of multidimensional citizenship in a knowledge space is discussed. With this in mind, the paper analyses the strategic role of natural language and conceptual modeling and presents some applications and lesson examples.

## 2. Education to Complexity

What do we mean by smart citizenship [15]? Is it about enjoying smart territory benefits or demanding innovative services? Does it concern drawbacks and dangers for freedom and privacy [16]? In all cases it involves being skilled enough to cope with modernity and new social instruments: education to complexity is paramount.

In a traditional context, citizens are usually able to recognize their position in a hierarchy (social or economic) and in a community. Nowadays, the challenge is to recognize this position in a network, as a node connected through a plurality of relations and processes, feedbacks and back interactions.

Before reflecting on what being a citizen in a complex context means, it is worth focusing on complexity and its consequences in social life.

According to Barabási conceptualization and language [17], a complex network is not just a matter of interconnected agents: complex systems display several organizing principles, which are, at some levels, encoded in their topology.

### 2.1. Small World

Small world propriety means that, despite their size, in most networks there is a relatively short path between any two nodes. The distance between two nodes is defined as the number of edges along the shortest connecting path: the most popular *small world* expression [18] is the "six degrees of separation" concept, proposed in 1967 by the social psychologist Stanley Milgram.

Small world property has inspired sociologists and politicians and it has often been declined in terms of interdependency and multiculturalism: in this paper, we would like to re-think it in terms of the citizens' right to short distances to services and information. Communities are nowadays experiencing an unprecedented situation: they are inside a highly interconnected external environment, even if the governance rules are still grounded on authority and accountability principles, often organized in terms of hierarchies. The gap between the hierarchy of responsibilities and the network of information is difficult to understand and accept.

The complexity of modern citizenship challenges calls for an original tradeoff between the capacity to exert full authority and power (through hierarchies) and the acknowledgement that only a network system of competencies, vigilance, and trust can assure transparency and democracy.

## 2.2. Clustering

A common property of complex networks is an inherent tendency to clustering, quantified by the clustering coefficient [19]: in most real networks, the clustering coefficient is typically much larger than it is in a random network of an equal number of nodes and edges.

Social networks provide practical evidence of this point. It is worth considering how clustering occurrences often happen without the cluster participants' knowledge: this is the case of those driven by marketing, political, or consensus reasons.

## 2.3. Degree Distribution

Not all nodes in a network have the same number of edges. Usually in a random graph, as the edges are placed randomly, the majority of nodes have approximately the same degree and their distribution $P(k)$, which gives rise to the probability that a randomly selected node has exactly $k$ edges, is a Poisson distribution. Empirical results show that for complex networks, the degree distribution significantly deviates from a Poisson distribution. Such networks are called scale-free [20]. A scale-free network originates when a node, having to establish a new association, prefers to link itself with a node already characterized by many connections, contributing to an exponential growth of its connections inside the network: in this way the functional form of $P(k)$ deviates from the Poisson distribution of $P(k)$ expected for a random graph.

A social consequence is that complex networks, also thanks to innovative technologies, indicate new leadership modalities sometimes outside the citizens' awareness or control.

## 3. Being a Citizen in a Smart and Complex Environment

Complex networks have strong implications in many domains, and there are many cases in which recognizing a system as such (from information, to epidemiology, to social systems, to communities) allows for having a deeper insight into its properties, potentialities, and behaviors [21], empowering, in this way, citizens' awareness. Adopting a network and complexity perspective for reflecting on citizenship may provide new insights in modern societies.

Citizenship has been one main narrative in describing construction strategies of political and social arrangements in modern societies [22]: complexity may have a deep impact on this.

After the 1949 Marshall's Cambridge Lecture [23], citizenship has been theorized not only as a legal and political status, but also as a social status [24], thus gaining a multidimensional character: the expansion of individual and collective rights, in fact, is traditionally attached to the status of citizen.

In modern societies, the process of rights claiming, which is a traditional active citizenship expression, is usually mediated by IT systems. Citizens ask for efficiency, accessibility, and performance; they demand an evolutive environment, able to recover from errors and disservices, oriented to ever-growing service provisions.

In smart communities, therefore, persons have a two-fold citizenship experience: on one side, the level of their rights recognition is satisfied as never before, on the other, services in support of rights compliance, being IT mediated, generate complex data networks, the digital territories previously outlined. The concepts of community and citizenship are deeply involved and changed: traditionally, a community has been modeled as a political or institutional unit, a set of personal and empirical relationships, or on the basis of the geographical spaces where lives happen. Nowadays a community is also a linguistic object, in which every citizen is both author and content of different narrations. When integrated together, they form the overall information system [25–27], which, according to the complexity paradigm previously discussed, could be extraordinarily more informative than the sum of its components.

The modern communities and citizenship dimensions previously outlined, fit very well with the smart city concept: despite the fact that the term "smart city" has become more and more widespread, its sense, definition and dimensions are not always clear [28,29]. As any city may be conceived in terms of relations and territories [30], in this paper smart city will be conceptualized as a particular context, capable of learning from feedbacks and reacting to its inhabitants needs: a territory in which information is used to shape relations useful for adapting services to its stakeholders' exigencies and rights.

Following the complex paradigm outlined above, we can see how *smartness* falls within this frame.

Being highly connected environments, smart cities present a network structure in which the encoded topological properties previously discussed are usually true: at what level is a citizen aware of being in a small world? Is he/she aware of the existence of a plurality of path to reach targets, information, institutions? Are citizens conscious of the existence of hubs? What kind of legitimacy do they have? In accordance with which process does a network edge become a hub?

In this perspective, the educational proposal for smart citizenship will be advanced and discussed [31].

### 3.1. Natural Language and Conceptual Modeling in Support of Citizenship

As previously outlined, it is widely acknowledged that living in a complex environment may provide advantages and potentialities. A question arises: are citizens also able to recognize and evaluate risks and costs connected to the services they are asking for?

Usually communities find a trade-off between benefits and costs linked to rights acknowledgment: for instance, in traditional social debates, there may be conflict about the costs of services delivery and their distribution. Nowadays we also have to take into account those linked to proper and legitimate management of the complex digital spaces generated by the acknowledged services. The risks associated with abuse are also to be considered.

A *smart* citizen, aware of being an element of a more general narration, must be provided with skills and instruments to shape digital spaces according to their visions, projects, rights protection.

As with any citizenship dimension, this awareness has to be granted to everyone, regardless of personal technical skills: in this paper, *natural language* and *conceptualizations* are the required competences: we are dealing with non-technical assets widely used by IT professionals to model and build smart services. The challenge is to also restore them to unskilled citizens in order to empower them and allow them to verify and discuss potential smart service risks for individuals and communities.

### 3.1.1. Natural Language

The first *tool* in support of smart citizenship is natural language: its role in information system design and construction is known and is related to careful and active listening to the users' needs and requirements.

It is not surprising that narrations are so strategic in both developing and using an information system: Jerome Bruner [32], for instance, has argued that one of the ways in which people understand their world is through the "narrative mode" of thought, which is concerned with human wants, needs, and goals.

In formal information system development, the production and collection of narrations and storytelling corresponds to the part of the design process in which the users' needs and requirements (along with their service vision) are collected and analyzed. It is worth noticing that this aspect in information system development is closer to traditional active citizenship: it concerns speaking about needs and the contexts and conditions for meeting them.

Information systems are usually developed in line with the methodology sketched in the following image: the boxes with red borders have, to a certain extent, been developed in natural language and, if properly arranged, they may even be up to non-skilled citizens.

Speaking of public services, for example, initialization could be a political phase in which citizen involvement is usually very high. System design may include dominion-experts: media and public opinion may demand information feedback. The figure shows that due to language and technical barriers, development and implementation are the only processes totally outside the citizens' comprehension.

### 3.1.2. Conceptual Models

The second tool is conceptual modeling. Any information system [33,34] is based on its conceptual model, produced on the basis of collected narrations and stakeholders' requirements.

The role of conceptual modeling in information system development is widely known: the value of conceptual models lies in their ability to capture the relevant knowledge about a domain, facilitating the engagement of involved stakeholders and supporting the reciprocal comprehension of users and designers.

Any real world narration is often translatable in a conceptual model, describing the semantics of the organization and representing assertions about its nature. Relations between ideas, images, or words are represented as well. Conceptual models may have a graphical representation, a map, in which concepts are usually enclosed in circles or boxes, and connecting lines between two concepts specify their relationships.

Conceptual models do not have any IT implications and support mutual comprehension between different languages and technical cultures: in other words, they cannot be considered a merely technological phase (even if they are compulsory in any information system design), as they are linked with the capability of properly describing and defining a problematic area.

The passage from narrations to conceptual models may be performed at different formality levels: in the educational experiences described and discussed in this paper the one closest to the natural language level has been chosen.

System designers use different modeling methodologies for (i) data (entity relationships, object oriented); (ii) processes (Idef0, FRAM); and (iii) interfaces (UML use cases). These approaches are all non-technical and natural language-based, and they are intended to bridge the gap between dominion experts and system developers.

It is always possible to harmonize these models: in the educational proposal, for instance, starting from a description of reality under the user services perspective, we intend to induce knowledge on the data system and on the processes, which generate and use these data.

### 3.2. Natural Language, Conceptual Models, and Data Spaces: A Bridge between System Users and Providers

In this paper, we advance the proposal that when a citizen, using the natural language, is able to recognize a service conceptual model, he/she is also capable of identifying databases, information repositories, informative potentialities as well as linking conditions with other sources: in this way, the *digital territory* created by the services used begins to take shape.

In this methodological proposal, conceptual models are adopted as an instrument per se, for their explanatory potentialities and their ability to orient unskilled citizens in digital territories. In this perspective, the adopted methodology relies on the use of conceptual mappings [35,36] as a support for education [37,38] to improve a meaningful understanding of the materials studied.

In this approach, concept maps met the need to explicitly show how new concepts and propositions were integrated into the learner's cognitive structure, helping students learn how to learn, capture explicit and tacit knowledge held by experts, assist in the design of instruction, facilitate creative work in every discipline, and facilitate improvement of management and marketing methods.

This proposal actualizes a pioneering use of conceptual mappings [39] and, in this perspective, it does not use conceptual models towards any form of operationalization, but only as an asset in support of comprehension and learning. An original methodological update to this consolidated educational perspective, is given by the fact that, in this case, conceptual modeling is specifically addressed to reconstruct and understand an IT service structure by its users' view, with the explicit objective of also learning about the potentialities and risks for privacy, freedom and full active citizenship expression.

A potential drawback is given by the fact that, usually, IT service conceptual models are not available for their users, who only approach them through their interfaces. In approach of this paper, we advance the proposal that this informative gap is bridged using the users' experiences and views developed in natural language as a source for a reverse conceptual modeling.

Citizens require services: their requests, expressed in natural language, are the basis for the new system structure and architecture. When constructed, citizens use the system and obtain the services requested without knowing how it has been built.

They are, in any case, able to describe the delivered services and what the system does: anyone is able to describe the experience of paying with a credit card, using an app to get a service, passing in a video-surveilled street. On the bases of these narrations, it is possible to derive and re-model a system conceptual model draft, useful for the citizens, to gain a clearer view of the digital space created by the services they use.

Moving from interfaces to system architecture and code is not an original approach and it is a classic example of reverse engineering. Citizenship empowerment does not need this technical specification level: it could be enough to broadly recognize the model on which the system architecture is based in its most abstract and general form.

Neither is the idea of using natural language to understand complex organization new [40]; it has quite a long tradition starting from problems of strategic management in complex environments. With respect to these early applications, technology has made important advances, even if the basic approach is still very similar: natural language is a useful modeling tool as it is rich enough to cope with the complexity of organizations.

To support these points, the following picture illustrates a very common situation: an institution wants to provide services to support a problematic area. In this perspective, information systems designers and producers build a problem/solution conceptual model, on the basis of which they create the system.

Citizens, as system users, are represented in the model: to provide the services, their data may be collected and integrated with other sources. They use the system through dedicated interfaces and often interfaces are their only perspective; the picture shows how a simplified system conceptual model draft may be derived from the users' point of view.

The users' conceptual model will not necessarily be the same as the one developed by system designers. It will be less detailed and the result of the joint efforts of different users profiles (illustrated in the picture with different colors), but it will still provide a deeper insight into the system's informative structure. Citizens will be more skilled in recognizing the data territory generated by their service usage.

The chart discussed and presented in Figure 1 has now been completed with the reverse modeling actions performed by citizens.

The following Table 1 summarizes the citizens' role in the different phases of an IT service development.

They are not condemned to transfer their sovereignty to technology: the Table 1 shows all the phases in which they could be actively involved.

Citizens (Figure 2) have several cards to play in order to achieve control and transparency, even in smart and intelligent communities, from the very beginning of the IT processes when they may claim participation in new service modeling, to the usage phase (Figure 3) in which they must be more skilled and conscious in order to be vigilant and critical of the effective costs of smart services.

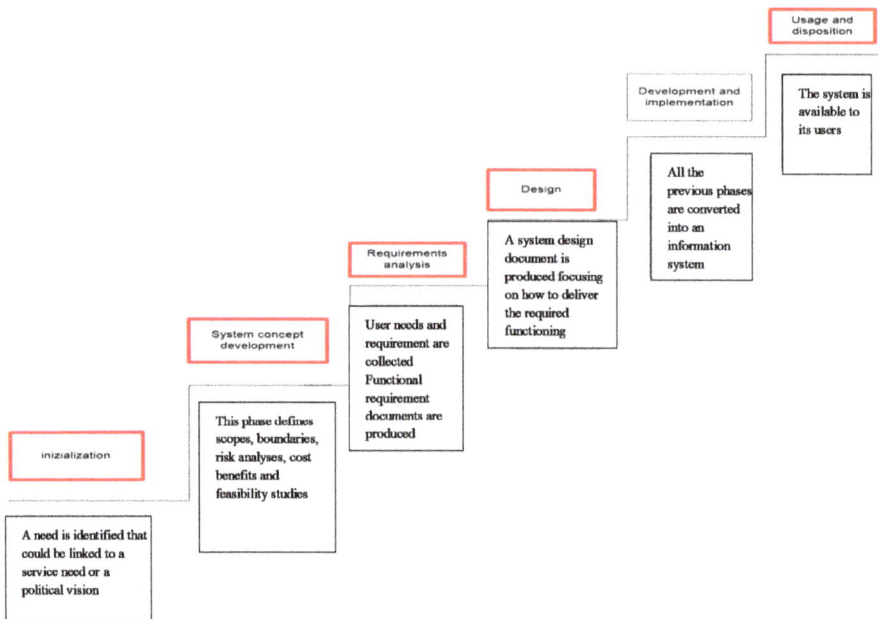

**Figure 1.** Natural language in the information system-developing scheme. With red borders are all of the phases affected by natural language.

**Table 1.** Role of citizens.

| Phase | Role | Process |
|---|---|---|
| Political decision on rights and services | active | Action on vision promotion and choice |
| Problematic area and description of services | active | Citizens describe their needs and requirements |
| | passive | System designers are collecting and organizing the narrations |
| Conceptual modeling of services | passive | Citizens are modeled |
| Construction of services | absent | Repositories for citizens' data are constructed |
| Service usage | active | The service is used and the right is granted |
| | passive | Data on citizen are collected, reused, integrated |

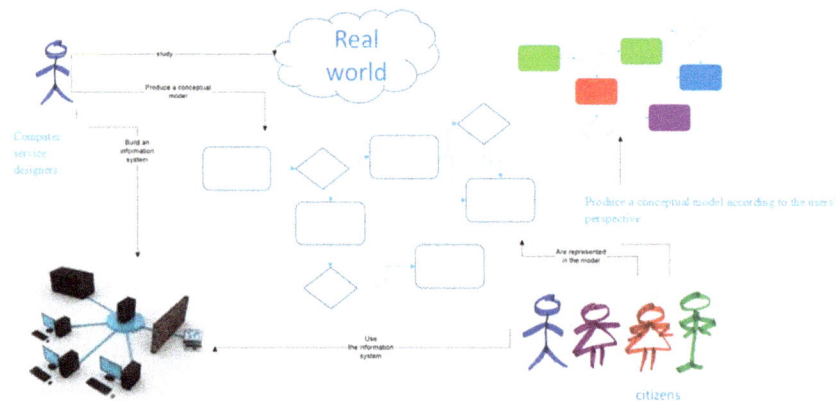

**Figure 2.** Smart services conceptual models as per the users' perspective.

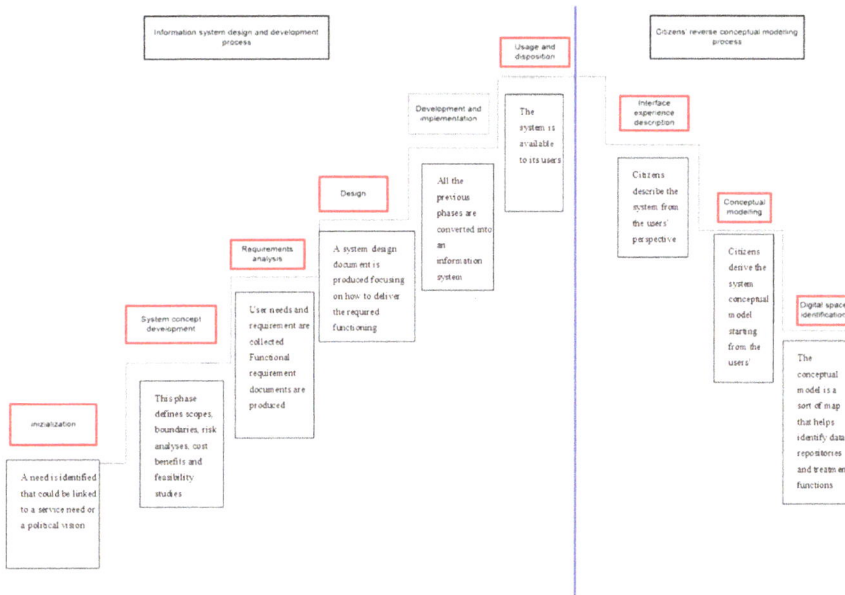

**Figure 3.** Citizens' reverse conceptual modeling process.

## 4. Smart Citizenship: Proposed Guidelines for the New Education Approach

The topics, previously discussed, have been developed in an educational proposal. Following are the didactical approach and the presentation of the experiences.

### 4.1. Making Teachers Work Together

To cope with new citizenship challenges, the educational system usually provides partial answers. Humanities teachers, for instance, could focus on laws and on the trade-off between people's rights to services and privacy. They could stimulate their students to extend to smart cities the attention usually reserved for press freedom and media propriety. IT teachers could focus on the technological aspects of instruments and services, possibly targeting innovation. Math and statistics teachers could focus on indicators and data analysis. All of these approaches may help, however they are narrow and deal with single issues.

If citizens are required to adapt to complexity, educational systems have to overcome the frequent lack of communication between subjects and disciplines. The objective is to identify a common ground to which every educational stakeholder contributes with specific languages and competencies for overall system comprehension. In line with the above, the required common ground is a conceptual model and the instrument is natural language.

### Unifying Subjects

Bridging the gap due to different languages, sectors and disciplines begins at school: the proposed educational methodology intends to support the co-working of teachers of different disciplines.

Often, in traditional education (Figure 4), the main focus is to ensure disciplinary consistency in terms of coherence, timing, and sequencing of the subject contents. The following picture shows this situation, in which subject outcomes are largely independent.

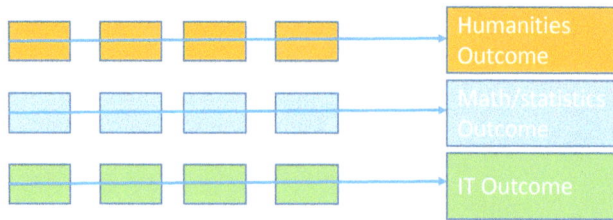

**Figure 4.** Traditional disciplinary content delivery.

It is widely known [41] that interdisciplinarity increases student engagement and helps them develop their cognitive abilities. "If establishing student learning outcomes for disciplinary courses and programs is hard, it may be even harder to achieve for interdisciplinary courses and programs where there is less agreement about what interdisciplinarity is, what outcomes should be assessed" [42].

The following chart illustrates the recommended teaching framework, in which learning outcomes are in part attributable to specific discipline objectives and, in part, to a more general and interdisciplinary vision.

The joint reference to the same conceptual model enhances parallel integration.

The following diagram shows how the interdisciplinary reference to the same conceptual model helps in bridging subject languages.

*4.2. The Method*

The proposed educational methodology starts with identification of a problematic area, which will be addressed by the joint effort of the different disciplines involved. On the basis of this choice, the following steps will be performed:

- Natural language reality description
- Transformation of this narrative into a set of short phrases (subject, verb, and predicate)
- Transformation of actors (the nouns in the narrative) into nodes
- Transformation of processes (the verbs in the narrative) into relationships
- Production of a graphical representation of this set of nodes and links
- Connection of node and relationship to existing data sets
- Analysis and description of the situation under investigation through quantitative and qualitative approaches.

The following Table 2 highlights the contribution of the disciplines and subjects involved:

**Table 2.** Disciplinary contributions and their outcomes.

| Subject | Action | Result |
|---------|--------|--------|
| Humanities | Narration production<br>Linguistic analysis<br>Conceptual model critical analysis | Texts<br>Conceptual model<br>Informative potentialities and drawbacks |
| Computer science | Data base design and conceptual model<br>Database construction<br>Tech innovation | Hypothesis of data structure<br>database<br>New sources, sensors, internet of things |
| Math and statistics | Production of indicator suited to the conceptual model<br>Data analysis based on the data collected by the service | Quantitative description of the problematic area |
| Humanities | Learning from data New insight into problems | |

The educational approach illustrated in the table is recursive and not sequential: Phase and roles may also be proposed in parallel to enhance the students' attitude to network reasoning.

### 4.2.1. The Role of Humanities Teachers

Humanities teachers do not normally perceive a specific role in technical education. In this approach, they make a major contribution as they are in charge of the narrative conceptualization.

After having chosen a particular topic, students, supported by their humanities teachers, produce a short paper describing, in natural language, a problematic area, typically a smart service.

To convert this narrative into a conceptual model, humanities teachers show how to transform a text into series of subject-predicate-object expressions, similar to classical conceptual modeling approaches such as entity-relationship or class diagrams. The texts, once conceptualized (i.e., with the entity/relationship approach), become maps and are available for a non-linear reading. While the original paper is made for linear reading, (left to right, start to finish), the maps can be navigated, like a hypertext: the sense is equivalent, the reading modalities are different. By working on the conceptual model, students are now able to explore the system's informative potentialities and to understand what happens when conceptual model elements are *filled* with data.

It must be noted that subject-predicate-object expressions allow for introducing, (when appropriate for the students age and level), elements of semantic and linguistic structural analysis that may be reused by math teachers (first order logic) and IT (semantic web).

### 4.2.2. The Role of Computer Science Teachers

Computer science teachers contribute to the program by introducing database elements, in compliance with the conceptual model. They discuss with the students the characteristics of the smart service data repositories, and check the availability of form of data restitution, such as open data or linked data.

They reflect on the technological strategies and implications of data linkage.

Sensors, the Internet of Things, and other innovative sources are presented and their role in the conceptual model is discussed.

Starting from the subject-predicate-object expressions produced with humanities teachers, if appropriate for the students' level, they could introduce semantic web elements as well as first-order logic elements together with math and statistics teachers. This aspect is very important as it introduces elements of logical inference on data, semantic web and big data. These elements, returned to the humanities teachers, are an interesting incentive to reflect on social consequences of the use of collected data, according to the structural approach discussed above.

### 4.2.3. The Role of Statistics and Math Teachers

A complex network contains topological properties that math teachers may present and discuss: the most important is the presence of clusters. Other data analysis aspects, in compliance with students' educational level, may also be introduced.

Statistical analysis results, once returned to the humanities teachers, may stimulate further reflections, in the perspective of the recursive process outlined in the previous paragraphs.

### 4.2.4. The Joint Role of All Teachers

Working on the same conceptual model to analyze the digital space is a good strategy for bridging different subjects. The aspects that may be addressed in a joint mode include:

- Linkage strategies, potentialities, and drawbacks between different sources and databases;
- The role of standard administration language in representing complex social networks;
- Grammar and first-order logic.

The approach previously discussed may be iterated, oriented, and investigated. The following diagram, which has been adapted from [41], reinstates the contents of Table 2 and shows a sort of spiral process [43] that involves different subjects and teachers.

The diagram shows how the problem is addressed by iterating the same methodology at different complexity levels.

The urgency of overcoming singularities in education in order to address knowledge production is obviously not new: this paper proposal has some similarities with the answer advanced by structuralism in the educational field. In that perspective, the concept of profession itself (such as the one of the teachers involved in Figures 5–7) has been deconstructed and reconceptualised as a *field* [44], assuming a role not far from the *conceptual model*: a common ground which is functional both to specialist insights and integrated judgments.

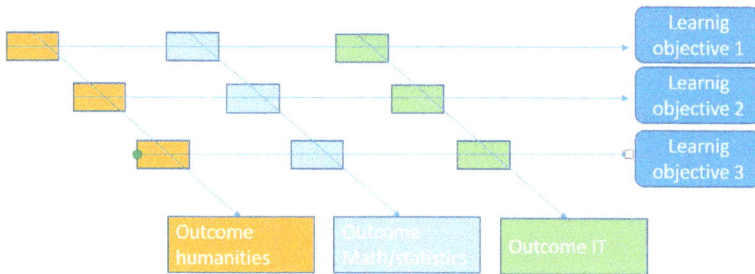

**Figure 5.** Interdisciplinary content delivery.

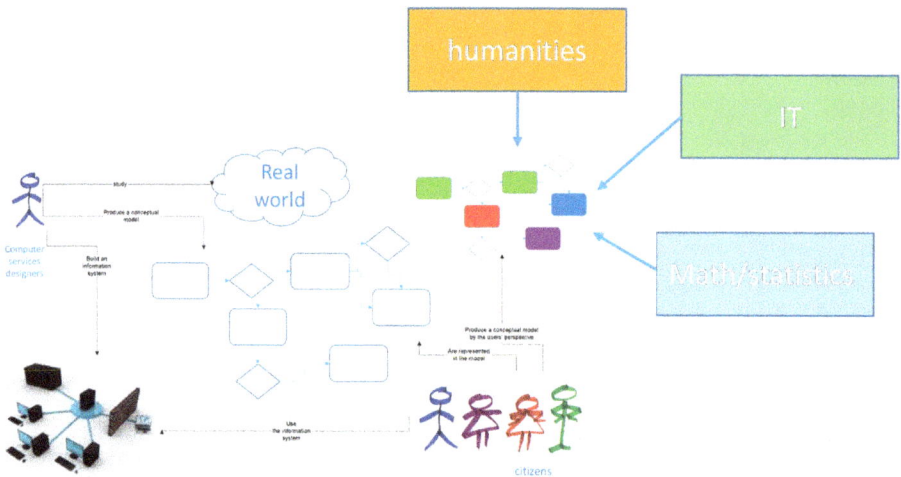

**Figure 6.** The contribution of conceptual models to an interdisciplinary education approach.

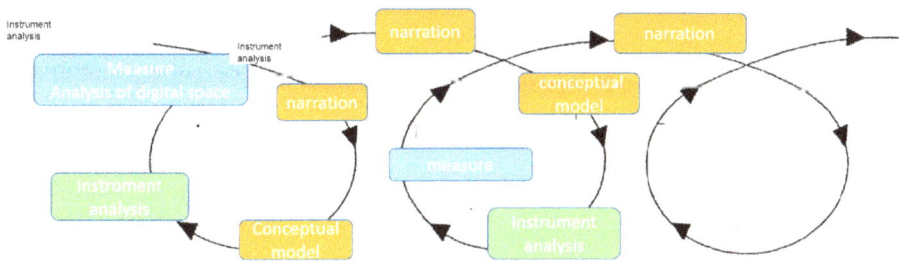

**Figure 7.** The iterative structure of the proposed methodology.

### 4.2.5. Towards a Revised Structuralist Approach?

It is not a coincidence that the concepts of *conceptual model*, the object of this proposal, and the one of *field* advanced by the structuralist approach are so close: underlying them both is the idea that a specific domain, like the digital space generated by smart services, may be understood mainly by means of a structure modeled on language.

Surrounding the idea of the sign and structure, structuralism developed the idea that everything that is acquired, transmitted, and shared is fused in a system that can be formalized as signs. The idea is, therefore, to analyze cultural reality as a language, or a system of signs. The sign, in turn, cannot be defined in an isolated manner. Its existence is relational. The sign's existence derives from the structure of the system, in which it is a knot or a point in the network.

In fact, structuralism [45] affirms that "elements of human culture must be understood in terms of their relationship with a larger, overarching system or structure. It works to uncover the structures that underlie all the things that humans do, think, perceive, and feel". According to Blackburn [46] nothing is intelligible "except through its interrelations. These relations constitute a structure, and behind local variations in the surface phenomena there are constant laws of abstract culture". Caution must be exercised when recognizing these parallels: nevertheless, with the structuralist and post-structuralist writers [47], the proposed educational approach is oriented to emphasize [48] the constitution of subjects and subjective experiences through discursive practices really related to structural signification. The post-structural framework, in particular, shows a high concern with language, its role in human praxis and the effects of narrations on praxis.

Structuralism has often been criticized for being unhistorical and favoring deterministic structural forces over the ability of individual people to act. These critical remarks were made before smart innovation and big data: nowadays they assume new relevance.

In a smart environment, the structure underlying the services (which are in turn the product of a narration) generates digital spaces on which postulations and deductions may automatically be performed. On this basis, private or public institutions may make decisions regarding citizens: this is what is actually happening in a semantic web fed by data generated by smart services. Traditional dichotomies of representation/reality or subjectivism/objectivism are probably overcome in smart territories: education has to address these new forms of knowledge in human praxis to protect social inclusion, and to tackle class reproduction and hierarchies of power, the opposite of the complex environment outlined at the beginning of this paper.

### 4.2.6. An Educational Scalable Approach

Scalability is an important dimension for an educational proposal addressing citizenship empowerment: scalability means that the approach may be adopted at different ages and carrier student phases by students, teachers, and normal citizens. The methodology remains the same.

The proposed approach has been tested in different ways: experiments have been performed with high school students, undergraduates, and graduates. Specific training courses have been proposed to teachers to help them to take advantage of the opportunities provided by the methodology: apart from differences in age and cultural level, the method was applied in the same way. This aspect has made cooperation possible among different groups as well as bridging the gap between different school levels.

*4.3. Heuristic Experiences of the New Educational Approach*

In the following, some of the examples proposed to students and teachers are presented and discussed.

4.3.1. Shopping with a Fidelity Card

One of the first topics proposed to students and teachers was entitled "What kind of information is generated by the use of a shopping fidelity card?"

Large retail sectors often propose fidelity cards to support loyalty programs. In marketing, a loyalty card identifies the cardholder as a member of a loyalty program. By presenting this card, purchasers usually earn the right either to a discount on the current purchase, or to an accumulation of points to be used for future purchases.

Figure 8 shows, as an example of the class work, one of the conceptual models produced on the basis of the narrations of the experience of the students as fidelity card users: the conceptual model produced allowed the students (16–18 years old) to examine loyalty program privacy concerns. In particular, humanities teachers discussed the informative implications in providing information when applying for cards, on family composition, profession, and personal preferences.

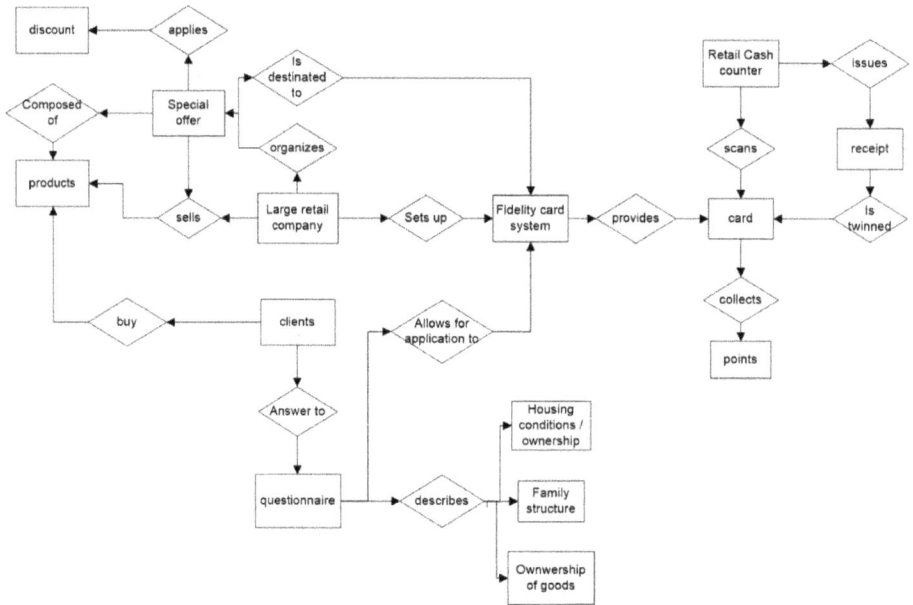

**Figure 8.** Using a fidelity card: the conceptual model.

The students also reflected on the fact that data could be available at disaggregated levels, even if stores usually apply agreements (typically non-disclosure) concerning customer privacy.

Companies, for instance, may use the information they gather to see to what extent they can modify consumer behavior. Some uses of shopping information are innocuous, but what if a company decides to sell or share personal data like the type of newspaper bought or regarding alcohol purchases?

In addition, how does one cope with aggressive marketing strategies, made possible by the joint knowledge of personal information and purchasing preferences?

Moreover, the conceptual modeling, allowing a nonlinear narration reading and a full exploration of the problematic area, helped in reflecting on the importance of a thorough reading of the privacy policies attached to shopping cards.

In this specific example, the model was produced by the class transforming the original narration into a list of short phrases made up of subjects, verbs (transitive, wherever possible) and objects. A simple semantic check was carried out to avoid synonyms and fuzzy concepts. The result was translated in the map.

In the educational program proposed, the conceptual model was also used by statistics and math teachers to introduce the indicators produced to sustain marketing policies and to profile purchasers' preferences. For the conceptual map elements, a reflection was stimulated on pertinent measures: different analytical perspectives were proposed and discussed. For instance: what kind of indicators can optimize the offer in relation to the demographic and social profile of cards owners? At what rate are specific goods bought by a certain purchaser cluster?

This further consideration allowed the class to reflect on the perspective, offered by the system, of verifying the correctness and accuracy of the original narrative, as the indicators may offer the chance to check the original assumptions. This is particularly important for reflecting on the importance of institutional statistical communication protocols in a transparency perspective.

### 4.3.2. Microchip Trash Collector in Support of Garbage Sorting

If in the example previously discussed, focus was placed on the relationships between private companies and their clients, a further example was provided on issues arising from the relationships with public service providers.

The students: (i) narrate; (ii) conceptualize; (iii) discuss; and (iv) measure and identify data concerning garbage-sorting policies.

Many communities adopt waste sorting policies in order to minimize residues and pollution. There is a wide consensus around these policies, but some issues regarding privacy may arise.

Adopting the same methodology described above, the students described their experience with regard to garbage sorting. They were then required to discuss and conceptualize some of the approaches used to verify the effective sorting practice implemented by families and citizens. In particular, they were required to model the situations in which the garbage can is identifiable. The model navigation exercise focused on the level of private information that could be derived from garbage inspection. The case in which only the penalized inspection results are recorded was modeled and discussed.

### 4.3.3. The Role of Standard Administration Language in Generating Complex Social Networks

In this lesson students were asked to identify potential data sources to fill and instancialize their conceptual models: this part is usually up to IT teachers. On this occasion the students also reflected on the importance of assuring a standard language in public administration. Government policy implementation is often hierarchically organized to ensure responsibility and accountability: services often involve several public bodies.

The diagram on the left (Figure 9) represents a service delivery structure: more specifically, the lesson focused on the administrative procedures following a work accident. The red elements represent, in the example, the management of workers' hospitalization and, on a completely different branch, the management office of legal affairs.

Students were asked to reflect on the importance of adopting the same administrative language in the red spots: a homogeneous language transforms the organizational hierarchy into an information network, characterized by the small-world property. The data collected in the two institutions are linkable and a knowledge network is generated. The system is able to connect legal and health elements and can teach more about causes, consequences, and responsibilities. Students reflected on the fact that administrative dialects may hamper data-mapping in the problematic area conceptual model and be a serious obstacle to transparency.

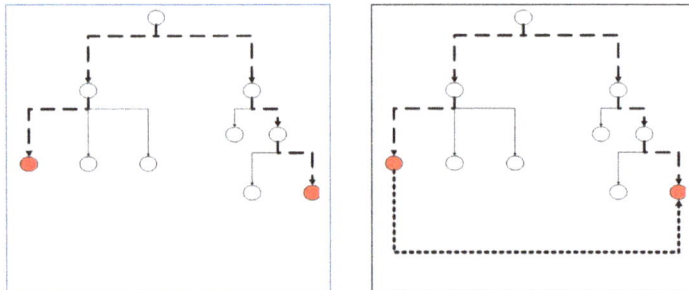

**Figure 9.** From the hierarchy of responsibilities to the knowledge network.

## 5. Final Remarks

The ever-growing demand for specialized smart services may have potentially dangerous side effects on freedom, privacy and, ultimately, on the concept itself of citizenship and democracy: with respect to this unprecedented situation there is a lack of awareness and knowledge that the education system ought to address, providing cultural assets, and empowering methods and attitudes.

Not an easy challenge, as it involves competencies and skills, which are often narrow and deal only with single issues: humanities, IT, math, and statistics teachers could stimulate and inform their students. However, if citizens are required to adapt to complexity, educational systems have to overcome the lack of communication between subjects and disciplines. The objective is to identify a common ground to which every educational stakeholder contributes with specific languages and competencies for overall comprehension.

This paper advances an educational proposal rooted on conceptual models of the users' experience, which are presented and taught not as the imperative prerequisite for information system operationalization, but as a general and non-technical skill to allow citizens to recognize, discuss and evaluate smart service informative structures.

In this perspective, conceptual models go beyond the IT area and become a crosscutting subject that bridges the gap between languages, sectors and disciplines, supporting the co-working of teachers of different disciplines.

Several examples of lessons are presented and discussed in this paper. The lectures were proposed to classes differing in age and level: the didactic methodology remained the same and was adapted to the students' age and profile with deeper reflections and insights.

In all the situations experienced, the students were able to formulate original ideas and reflections on the potential costs of the citizens' right to avail of smart and intelligent services, using the conceptual models produced as a map to be explored, such as unwanted connections with data sources outside the analyzed smart service, for instance.

Beyond the promising results achieved, a major critical factor may be attributed to the teachers' specific sensibility and preparation on the topic: for this reason, the didactical proposals have to be formulated in a two-step structure: an action addressing educators (together with proper supporting material) and a specific step targeting the students.

It is a complex educational goal: the successfully perceived outcomes of such heuristic experiences indicate further improvement and development directions and motivate us to work out a systematic approach for application of the proposed education guidelines, as well as for its effective evaluation and validation.

**Conflicts of Interest:** The author declares no conflict of interest.

## References

1.  Bargh, J.A.; McKenna, K.Y. The Internet and social life. *Annu. Rev. Psychol.* **2004**, *55*, 573–590. [CrossRef] [PubMed]
2.  Nam, T.; Pardo, T.A. Conceptualizing Smart City with Dimensions of Technology, People, and Institutions. In Proceedings of the 12th Annual International Digital Government Research Conference: Digital Government Innovation in Challenging Times, College Park, MD, USA, 12–15 June 2011; pp. 282–291.
3.  Hirsch, E.; Silverstone, R. *Consuming Technologies: Media and Information in Domestic Spaces*; Routledge: Abingdon-on-Thames, UK, 2003.
4.  Murthy, D. Digital ethnography an examination of the use of new technologies for social research. *Sociology* **2008**, *42*, 837–855. [CrossRef]
5.  Kraut, R.E.; Fussell, S.R.; Brennan, S.E.; Siegel, J. Understanding effects of proximity on collaboration: Implications for technologies to support remote collaborative work. *Distrib. Work* **2002**, *1*, 137–162.
6.  Trainor, K.J.; Andzulis, J.M.; Rapp, A.; Agnihotri, R. Social media technology usage and customer relationship performance: A capabilities-based examination of social CRM. *J. Bus. Res.* **2014**, *67*, 1201–1208. [CrossRef]
7.  Jarvis, P. *Democracy, Lifelong Learning and the Learning Society: Active Citizenship in a Late Modern Age*; Routledge: Abingdon-on-Thames, UK, 2008.
8.  Turner, B.S. Outline of a Theory of Citizenship. *Sociology* **1990**, *24*, 189–217. [CrossRef]
9.  Audigier, F. *Basic Concepts and Core Competencies for Education for Democratic Citizenship*; Council of Europe: Strasbourg, France, 2000.
10. Rheingold, H. *Smart MOBS: The Next Social Revolution*; Basic Books: New York, NY, USA, 2007.
11. Gaonkar, S.; Li, J.; Choudhury, R.R.; Cox, L.; Schmidt, A. Micro-blog: Sharing and Querying Content through Mobile Phones and Social Participation. In Proceedings of the 6th international conference on Mobile systems, applications, and Services, Breckenridge, CO, USA, 17–20 June 2008.
12. Palen, L.; Anderson, K.M.; Mark, J.; Martin, J.; Sicker, D.; Palmer, M.; Grunwald, D. A Vision for Technology-Mediated Support for Public Participation & Assistance in Mass Emergencies and Disasters. In Proceedings of the 2010 ACM-BCS Visions of Computer Science Conference, Edinburgh, UK, 13–16 April 2010.
13. Haste, H. Constructing the citizen. *Political Psychol.* **2004**, *25*, 413–439. [CrossRef]
14. Bennett, W.L.; Welles, C.; Rank, A. Young citizens and civic learning: Two paradigms of citizenship in the digital age. *Citizensh. Stud.* **2009**, *13*, 105–120. [CrossRef]
15. Mossberger, K.; Tolbert, C.J.; McNeal, R.S. *Digital Citizenship: The Internet, Society, and Participation*; MIT Press: Cambridge, MA, USA, 2007.
16. Bovens, M. Information rights: Citizenship in the information society. *J. Political Philos.* **2002**, *10*, 317–341. [CrossRef]
17. Albert, R.; Barabási, A.L. Statistical mechanics of complex networks. *Rev. Mod. Phys.* **2002**, *74*, 47. [CrossRef]
18. Milgram, S. The small world problem. *Psychol. Today* **1967**, *2*, 60–67.
19. Watts, D.J.; Strogatz, S.H. Collective dynamics of 'small-world' networks. *Nature* **1998**, *393*, 440–442. [CrossRef] [PubMed]
20. Barabási, A.L.; Albert, R. Emergence of scaling in random networks. *Science* **1999**, *286*, 509–512. [PubMed]
21. Fogelman-Soulié, F. Evolving Networks. In *Mining Massive Data Sets for Security: Advances in Data Mining, Search, Social Networks and Text. Mining, and Their Applications to Security*; IOS Press: Amsterdam, The Netherlands, 2008; p. 19.
22. Janoski, T. Conflict, Citizenship and Civil Society. *Contemp. Sociol. J. Rev.* **2011**, *40*, 151–153. [CrossRef]
23. Turner, B.S. TH Marshall, social rights and English national identity: Thinking Citizenship Series. *Citizensh. Stud.* **2009**, *13*, 65–73. [CrossRef]
24. Marshall, T.H. *Citizenship and Social Class*; Cambridge University Press: Cambridge, UK, 1950; Volume 11.
25. Agranoff, R. Inside collaborative networks: Ten lessons for public managers. *Public Adm. Rev.* **2006**, *66*, 56–65. [CrossRef]
26. Börzel, T.A. Organizing Babylon. On the different conceptions of policy networks. *Public Adm.* **1998**, *76*, 253–273. [CrossRef]

27. Macklin, T.; Jenket, P. Achieving Cross-Domain Collaboration. In *RTO IST Symposium on Coalition C4ISR Architectures and Information Exchange Capabilities*; NATO: Hague, The Netherlands, 2004.

28. Chourabi, H.; Nam, T.; Walker, S.; Gil-Garcia, J.R.; Mellouli, S.; Nahon, K.; Scholl, H.J. Understanding Smart Cities: An Integrative Framework. In Proceedings of the 2012 45th Hawaii International Conference on System Science (HICSS), Maui, HI, USA, 4–7 January 2012.

29. Schaffers, H.; Komninos, N.; Pallot, M.; Trousse, B.; Nilsson, M.; Oliveira, A. Smart Cities and the Future Internet: Towards Cooperation Frameworks for Open Innovation. In *The Future Internet Assembly*; Springer: Berlin/Heidelberg, Germany, 2011.

30. McCann, E.; Ward, K. Relationality/territoriality: Toward a conceptualization of cities in the world. *Geoforum* **2010**, *41*, 175–184. [CrossRef]

31. Martelli, C. A Linguistic Approach to the Construction of Complex Statistical Information Systems. In *Survey and Statistical Computing IV. The Impact of Technology on the Survey Process*; Birks, D., Banks, R., Gerrard, L., Johnson, A.J., Khan, R., Macer, T., Wills, P., Eds.; Association for Survey Computing: New York, NY, USA, 2011; p. 185.

32. Bruner, J. *Acts of Meaning*; Harvard University Press: Cambridge, MA, USA, 1990.

33. Wand, Y.; Monachi, D.E.; Parsons, J.; Woo, C.C. Theoretical foundations for conceptual modelling in information systems development. *Decis. Support Syst.* **1995**, *15*, 285–304. [CrossRef]

34. Avison, D.; Fitzgerald, G. *Information Systems Development: Methodologies, Techniques and Tools*; McGraw Hill: New York, NY, USA, 2003.

35. Novak, J.D. *Learning, Creating, and Using Knowledge: Concept Maps as Facilitative Tools in Schools and Corporations*; Routledge: Abingdon-on-Thames, UK, 2010.

36. Novak, J.D.; Cañas, A.J. The Theory Underlying Concept Maps and How to Construct and Use Them. Available online: http://eprint.ihmc.us/5/ (accessed on 15 June 2016).

37. Hartley, R. Conceptualizing and supporting the learning process by conceptual mapping. *Smart Learn. Environ.* **2014**, *1*, 1. [CrossRef]

38. Martin, L.G. A New Model for Adult Literacy Education: Technology-Based Concept Mapping in GED Preparation. Available online: http://newprairiepress.org/aerc/2014/papers/50 (accessed on 20 January 2017).

39. Novak, J.D.; Gowin, D.B. *Learning How to Learn*; Cambridge University Press: Cambridge, UK, 1984.

40. Daft, R.L.; Wiginton, J.C. Language and organization. *Acad. Manag. Rev.* **1979**, *4*, 179–191.

41. Carrington, S.; Robinson, R. A case study of inclusive school development: A journey of learning. *Int. J. Incl. Educ.* **2004**, *8*, 141–153. [CrossRef]

42. Repko, A.F. Transforming an Experimental Innovation into a Sustainable Academic Program at the University of Texas-Arlington. In *The Politics of Interdisciplinary Studies: Essays on Transformations in American Undergraduate Programs*; Augsburg, T., Henry, S., Eds.; McFarland: Jefferson, NC, USA, 2009.

43. Etzkowitz, H.; Ranga, M. A Triple Helix System for Knowledge-Based Regional Development: From "Spheres" to "Spaces". In Proceedings of the VIII Triple Helix Conference, Madrid, Spain, 20–22 October 2010.

44. Naidoo, R. Fields and institutional strategy: Bourdieu on the relationship between higher education, inequality and society. *Br. J. Sociol. Educ.* **2004**, *25*, 457–471. [CrossRef]

45. Munawar, B.; Rafique, H. Structuralist Analysis of the Poem "The Stone Chat" by Taufeeq Rafat in perspective of Binary Opposition. *Int. J. Appl. Linguist. English Lit.* **2016**, *5*, 122–126.

46. Blackburn, S. *Oxford Dictionary of Philosophy*, 2nd ed.; Oxford University Press: Oxford, UK, 2008.

47. Grenfell, M. Bourdieu and Initial Teacher Education-a post-structuralist approach. *Br. Educ. Res. J.* **1996**, *22*, 287–303. [CrossRef]

48. Edwards, R.; Usher, R. *Postmodernism and Education: Different Voices, Different Worlds*; Routledge: Abingdon-on-Thames, UK, 2002.

![future internet logo] *future internet*

[MDPI logo]

*Article*

# A Service-Oriented Approach for Dynamic Chaining of Virtual Network Functions over Multi-Provider Software-Defined Networks

**Barbara Martini** [1,*] **and Federica Paganelli** [2,*]

[1]   National Interuniversity Consortium for Telecommunications (CNIT),
      National Laboratory of Photonic Networks, via G. Moruzzi 1, 56124 Pisa, Italy
[2]   National Interuniversity Consortium for Telecommunications (CNIT),
      Research Unit at the University of Firenze, via S. Marta 3, 50139 Firenze, Italy
*    Correspondence: barbara.martini@cnit.it (B.M.); federica.paganelli@cnit.it (F.P.); Tel.: +39-050-5492245 (B.M.);
     +39-055-2758524 (F.P.)

Academic Editor: Dino Giuli
Received: 23 February 2016; Accepted: 18 May 2016; Published: 1 June 2016

**Abstract:** Emerging technologies such as Software-Defined Networks (SDN) and Network Function Virtualization (NFV) promise to address cost reduction and flexibility in network operation while enabling innovative network service delivery models. However, operational network service delivery solutions still need to be developed that actually exploit these technologies, especially at the multi-provider level. Indeed, the implementation of network functions as software running over a virtualized infrastructure and provisioned on a service basis let one envisage an ecosystem of network services that are dynamically and flexibly assembled by orchestrating Virtual Network Functions even across different provider domains, thereby coping with changeable user and service requirements and context conditions. In this paper we propose an approach that adopts Service-Oriented Architecture (SOA) technology-agnostic architectural guidelines in the design of a solution for orchestrating and dynamically chaining Virtual Network Functions. We discuss how SOA, NFV, and SDN may complement each other in realizing dynamic network function chaining through service composition specification, service selection, service delivery, and placement tasks. Then, we describe the architecture of a SOA-inspired NFV orchestrator, which leverages SDN-based network control capabilities to address an effective delivery of elastic chains of Virtual Network Functions. Preliminary results of prototype implementation and testing activities are also presented. The benefits for Network Service Providers are also described that derive from the adaptive network service provisioning in a multi-provider environment through the orchestration of computing and networking services to provide end users with an enhanced service experience.

**Keywords:** network function virtualization; software defined networking; service-oriented architecture; network service; service chaining; virtual network function

---

## 1. Introduction

Today's users continuously demand interactive, content-rich, and immersive networked experiences while imposing increasingly stringent requirements on the delivery capability of the network infrastructure. With a number of mobile digital services, *i.e.*, apps, which are racing toward the two million mark, the over-the-top (OTT) service providers have a consistent margin for innovating their infrastructure to optimize service provisioning and to enrich their service offerings to users at a rapid pace. On the other hand, network service providers (NSPs) have many difficulties to roll out innovative network-based services and to benefit from increasing revenues of the new digital economy [1]. A relevant obstacle is the inflexibility of the network infrastructure, which is static and costly to change. In fact, the deployment of network services and functions (e.g., routers, middleboxes) traditionally require the acquisition and operation of specialized hardware devices and their interconnections. This results in static chains of network services that cannot flexibly cope with dynamic user and service requirements (e.g., delay constraints) [2]. Moreover, the inflexibility in the service delivery is even more prominent across different NSPs and geographical domains where the level of resource accessibility and exploitation in the service delivery chain is still limited. The lack of a multi-provider infrastructure service coordination and deployments prevents NSPs to benefit from agile service customization, service enhancement, and reaching new markets [3].

Recent research efforts on promising network technologies, *i.e.*, software defined networking (SDN) [4] and network function virtualization (NFV) [5], go in the direction of conceiving and developing novel network virtualization models and programmable network service delivery functions to address the aforementioned flexibility and scalability requirements. Indeed, in accordance with recent service virtualization paradigms adopted in the computing domain (*i.e.*, cloud computing), they are expected to promote a service-centric vision where the data delivery function is decoupled from the underlying network physical infrastructure and is conceived as a chain of ready-to-use functional capabilities, dynamically deployed where more appropriate in the physical network as virtual resources and provisioned as a service in potentially multi-provider environments [6,7]. Challenges related to dynamic service composition and provisioning in a multi-provider environment have been faced in the last two decades in the IT domain, especially with the efforts related to service-oriented architecture (SOA) [8]. We, thus, deem it important to take into account principles and best practices of SOA in order to foster the deployment of SDN and NFV solutions for the provision of network services that can also be dynamically discovered, negotiated, and composed from different providers to meet specific user and service requirements.

Below, we introduce the main concepts of NFV and SDN in order to provide the background information for our work. We also introduce main concepts of SOA, since it provides a reference framework for service modeling and provisioning.

*1.1. Background*

Network Function Virtualization [5] is one of the most innovative manifestations of virtualization in networking enabling network functions and capabilities to be implemented as software components and executed in virtual machines or containers provisioned in general-purpose hardware systems, *i.e.*, VNFs. Thanks to a more agile lifecycle management of virtual machines, VNFs can be instantiated, updated, deleted when and where needed, as well as dynamically combined to provide more complex capabilities on demand. This allows for a strong reduction of capacity over-provisioning and for an opportunistic deployment and/or re-arrangement of VNFs to be performed, thus optimizing the usage of the underlying resource infrastructure toward different management targets (e.g., consolidated or balanced usage of physical servers, proximity to users).

Software-Defined Networking [4] is a new approach to the design, building, and management of networks. Basically, SDN decouples the software-based control plane (e.g., routing decision functions) from the hardware-based data plane (*i.e.*, packet forwarding engine) while abstracting the underlying network infrastructure and moving the network intelligence to a centralized software-based controller where network services, such as traffic engineering and path provisioning, are deployed. Such separation allows for a more agile and cost-effective network operation thanks to full programmability of forwarding capabilities and enhanced decision-making capabilities based on a global view of the network status. As a result, SDN opens the way toward a more effective interaction between applications and networks for the establishment of data delivery paths while addressing resource usage optimization and reducing the complexity of the network operation [4].

Service Oriented Architecture [8] is an architectural paradigm for the interoperability of heterogeneous systems from different administrative domains. Basically, service orientation principles define how resources and capabilities can be handled as independent services that can be flexibly and dynamically composed to provide more complex functionalities and address dynamic requirements. A service can be defined as a component that performs a simple, granular, and self-contained function that can be invoked by external clients through well-defined interfaces.

Erl [8] proposed the following main principles for service oriented design: service contract, loose coupling, service abstraction, service reusability, and service composability. The service contract expresses the capabilities offered by services and their technical interface details. The loose coupling principle emphasizes the need for reducing dependencies among the service implementation, its published contract, and service consumers. Service abstraction is a cross-cutting aspect of service-orientation that consists in hiding as much as possible the low-level details of a service interface and implementation. Service Reusability implies that services should be designed for serving more than one consumer. The Service Composability principle expresses the need for services that can be used as building blocks of more complex services (*i.e.*, composite services).

A SOA is typically characterized by three main roles: (i) the Service Provider makes the service available and publishes its description profile (contract) in a service registry handled by a Service Broker; (ii) the Service Consumer uses the service; and (iii) the Service Broker mediates the interactions between Service Providers and Consumers by offering discovery, matchmaking, and composition capabilities.

### 1.2. Contribution

SOA principles have been adopted in the past for the rapid and flexible development and management of value-added telecom services. However, a literature review on SOA models for Next Generation Networks [9] shows that, while many works focused on the integration of distributed service components at the service stratum, the benefits of applying the SOA paradigm to the transport functionalities have been scarcely investigated. Furthermore, recent works on network and service virtualization [10] show that the opportunity to adopt the SOA approach in telecom cloud environments remains quite unexplored. We argue that the adoption of SOA principles in the NFV and SDN technological landscape can ease the provision of network services that can be dynamically discovered, negotiated, and composed also from different providers to meet specific user and service requirements [7,11].

In this work, we propose a service-oriented approach for the orchestration of network services deployed as a cloud of Virtual Network Functions (VNF) on top of a NFV infrastructure that also takes full advantage of granular traffic steering capabilities provided by SDN. To this purpose, we refer to the SOA technology-agnostic architectural guidelines for organizing, (re)using, and integrating distributed networking capabilities provided by different systems [8]. First, we provide a survey of reference service scenarios for virtualized networks and, then, we elaborate how SOA, NFV, and SDN provide complementary features toward adaptive composite network service delivery, and organize these features in a set of tasks required for realizing dynamic network service chaining: service composition specification, service selection, service delivery, and placement. Then, we describe the architectural design of a SOA-inspired NFV orchestrator with SDN network control capabilities, also providing references to related standardization initiatives and illustrating its main workflows. We also present preliminary results of our ongoing activities for the implementation and testing of a prototype to validate the proposed approach. Finally, we conclude the paper with a discussion of benefits for NSPs along with a description of research challenges posed by the proposed approach.

## 2. Related Work

In this section we first provide an overview of main standardization activities in the NFV and SDN domains. Then, we discuss research works on dynamic network service chaining, highlighting contributions in terms of architectural and methodological guidelines and, finally, we motivate the contribution of our work.

*2.1. Standardization Activities*

The NFV ISG (Industry Specification Group) of the European Telecommunications Standard Institute (ETSI) is the most noteworthy standardization initiative so far regarding the NFV domain aiming at specifying general, open, and scalable architectural and operation solutions to meet challenges placed by NFV. More specifically, current standardization efforts in ETSI [5] are paving the way toward the realization of a multi-provider NFV ecosystem through the definition of a comprehensive architectural framework composed of: (i) the NFV Infrastructure (NFVI), which includes the hardware and software resources that support a cloud of virtualized network, computational and storage resources; (ii) the VNFs, which refer to the software implementations of network functions capable of running over the NFVI; and (iii) NFV Management and Orchestration (MANO), which covers the lifecycle management of both physical and software resources as well as management and orchestration of VNF instances [12]. An important step for NFV MANO has been to include software-defined networking solutions into the NFV architectural framework. Such solutions consider SDN controllers acting as Network Controllers for the network part of the NFVI (*i.e.*, network infrastructure domains) to deliver connectivity services involving physical or virtual resources (*i.e.*, OpenFlow switches) while providing an abstract view of such resources for orchestration purposes [13]. Moreover, in [14] ETSI identified the most common design patterns for using SDN in an NFV architectural framework along with recommendations to be fulfilled by the entities that perform the integration.

Current standardization efforts in the Internet Engineering Task Force (IETF) around SDN and NFV include, firstly, the Service Function Chaining (SFC) Working Group [15] efforts aiming at addressing the dynamic specification and instantiation of an ordered list of instances of service functions (such as firewalls, load balancers, *etc.*) while subsequently steering of data traffic flows through those service functions, accordingly. Secondly, another related initiative carried out in IRTF is the Network Function Virtualization Research Group (NFVRG) [16] aiming at developing new architectures, systems, and software, and to explore trade-offs and possibilities for leveraging virtualized infrastructure to provide support for network functions. Thirdly, efforts in IETF are also devoted to the design of a comprehensive SDN controller, namely Application-Based Network Operations (ABNO) architecture [17], including instrumental guidelines and operational workflow specifications to coordinate network control functions to compute paths, enforce policies and manage topology while providing a full network automation and programmability for the benefit of the applications that use the network. To the best of our knowledge, these efforts are still poorly integrated.

IEEE with the Next Generation Service Overlay Network (NGSON) standard [18] specifies a framework for the control and delivery of composite services over diverse IP-based networks (e.g., Internet, P2P overlay, IMS, PSTN, Mobile) with context-aware, dynamically-adaptive, and self-organizing capabilities. Dynamic context of users, devices, services, and networks are considered to adapt composite service delivery while optimizing network and computing resources consumption. In particular, network awareness allows for adapting service provisioning to the current status of the network (e.g., avoiding hot spots or congestions) while satisfying the ever-charging requirements of users. In this regard, recent emerging paradigms such as SDN and initiatives such as NFV can significantly contribute to achieve greater elasticity in network service deployments and accelerate a prominent innovation in NGSON service and data delivery functions. In this regard, a liaison has been established between the IEEE SDN Initiative and the NGSON WG to identify potential new standards in SDN/NFV areas to be developed in IEEE [19]. "The goals are to accelerate the proliferation of SDN services and applications and to offer a more efficient way of providing them through a service-architecture ecosystem of one-stop shopping for service-specific challenges" [20]. In [11] the authors report on their contributions to IEEE NGSON architecture to include SDN and NFV technologies and to provide a more powerful service composition and orchestration functions through generalized service chains including both application and network services while dynamically establishing data delivery paths across services.

The Open Networking Foundation (ONF) is a user-driven organization dedicated to the promotion and adoption of SDN through the development of the OpenFlow specifications as an open standard for the communication between the controller and the data forwarding network elements. ONF is also devoted to the promotion of open source developments as a way of consolidating impacts of SDN in the Industry. Moreover, ONF is very active in the definition of an SDN architecture framework with focus on the controller capabilities and on the interfaces with the other elements in the architecture both at north-bound and south-bound with applications and network elements, respectively [21,22]. Within the services area, an initiative started about providing an end-to-end orchestration, abstraction and resource optimization across data center SDN controllers and Wide Area Network (WAN) SDN controllers so that user-applications can be created and managed seamlessly [23]. In the NFV arena, the ONF has proposed a flexible NFV networking solution [24] for an NFV deployment of an OpenFlow-enabled SDN approach to deal with the dynamic provisioning of networking services. However, both initiatives do not cover the dynamic composition and orchestration aspects as this work does.

Other related standardization efforts are in progress in the Broadband Forum (BBF) [25], which is working on how cloud-based technologies including NFV can be used in the implementation of the Multi Service Broadband Network. The TMForum (TMF) rolled out the Zero-touch Orchestration, Operations and Management (ZOOM) program to develop best practices and standards to deliver true business agility and new digital services and revenue opportunities in the area of virtualization, NFV, and SDN [26].

## 2.2. State of the Art of Research Works

Dynamic service chaining is considered a relevant challenge for the evolution of network deployments toward flexible, cost-effective, and on-demand service delivery models. Recently, several works have been focused on this topic and related issues. Based on the analysis of existing works, we distinguish related literature in two main areas: architectural approaches for NFV/SDN integrated management; algorithms and techniques for specific issues in dynamic service chaining.

Within the first area, several works include an architectural model integrating NFV management with SDN control capabilities as a main goal of the work or as a reference framework within which more focused contributions can be positioned.

Some authors limited their discussion to high-level guidelines and architectural views. Reference [27] is a position paper that analyses the case of multi-domain distributed deployment of NFV. Some reference use cases are analyzed and challenges and research directions are discussed. Lopez *et al.* [28] argue the need of conceiving a Network Operating System (NOS) to cope with the lack of network-wide abstractions, thus favoring the foundation for true network programmability. However, this is a short paper that not handles in detail the integration of network control capabilities with NFV management functions. Naudts *et al.* [29] present a three-layer architectural view for both network and cloud domains (infrastructure, control and application layers). The integration of these two architectures is discussed by introducing an orchestration functional box, which is used for services that require a combination of these resources. However, its interaction with cloud and network control functions is only briefly described. Reference [30] analyzes benefits and challenges related to the adoption of the NFV paradigm toward the 5G cellular framework. The work moves from the high-level architecture conceived within the T-NOVA European Project to support the dynamic provision of VNFs on-demand and as-a-service. However, the authors provide only an introductory and conceptual description of the architecture.

Other related works provide deeper insights into architectural design and related implementation issues [31–34].

Reference [31] presents a functional architecture supporting automated, dynamic service creation leveraging NFV, SDN, and cloud virtualization techniques, which has been conceived in the framework of UNIFY, a three year European Project that has just concluded. The proposed architecture comprises a service layer, an orchestration layer and an infrastructure layer. They also briefly discuss how some functions/layers of the architecture can be mapped to ETSI NFV MANO and ONF SDN architectural elements. UNIFY do not explicitly refer to SOA principles, although some similar concepts exists, such as the description of services at different levels of abstractions (*i.e.*, service graphs and network function forwarding graph). Nevertheless, UNIFY does not aim at supporting a multi provider marketplace of network services as this work proposes using SOA practices, which were, in fact, designed to ease the operation of multi-provider environments where services can be dynamically-advertised and discovered. Giotis *et al.* [32] propose a modular VNF architecture providing policy-based management of VNF and service chains. Their main contribution consists in a basic ontology-based information model to describe network resources, network control functions, and VNFs capabilities with a uniform language. However, the role of network control functions (*i.e.*, SDN controller) in the architecture and its interaction with the NFV orchestrator are not explained. Munoz *et al.* [33] address the problem of SDN-based virtual connectivity over multi-technological domains. Therefore they propose an integrated SDN/NFV management and orchestration architecture which is specifically conceived for the dynamic deployment of Virtual Tenant Networks and the related SDN Controllers (implemented as VNF in DCs). However, they do not address the service chaining problem, as this works does.

Soares *et al.* [34] propose the CloudNFV Platform for enabling a telecom operator to deploy and manage service functions in a distributed cloud infrastructure. Although they focus on service chaining, their work differs from our contribution in that the proposed architecture specifically reflects the software design of their prototype (*i.e.*, instead of providing general functional specifications, some functional blocks are defined in terms of software implementation OpenStack and OpenDayLight). Moreover, they do not explicitly refer to service oriented principles and architectural guidelines.

On the other side, Garay *et al.* [35] stress the importance of a description model for network service. Although they do not explicitly refer to SOA guidelines, this approach is in principle compliant with those guidelines. However, their contribution consists in a straw man model for service descriptions, not on architectural frameworks.

As regards the second area of research we identified several authors focusing on the problem of VNF placement, *i.e.*, the (sub)optimal placing of a set of virtual network functions on a network of physical nodes to serve a set of service chain requests while optimizing a certain utility function, for instance to minimize the number of utilized servers [36,37] or to apply load balancing policies [38]. In [39] the authors provide a formalization of the VNF scheduling problem, *i.e.*, finding the corresponding time slots for functions to be executed over a given set of machines. Other authors [40] focused on the routing problem, *i.e.*, assigning paths to the incoming traffic flow requests in order to connect nodes running virtual functions. Ultimately, such works focused on specific issues of service chaining, proposing focused algorithms and techniques, but, to the best of our knowledge none of them discuss how the proposed solution fits into a comprehensive architectural framework.

### 2.3. Motivation of Our Work

The analysis of the literature confirms the need of reference architectural frameworks supporting the integration of VNF orchestration and network control functions toward the dynamic provisioning of network services in multi-provider environments.

The main limitation of related work is the level of abstraction adopted for the specification of the reference architectural framework: some works provide high-level and conceptual guidelines [27–30], while others present architectural design which are coupled with implementation choices [33,34].

Our objective is, thus, to propose a functional architecture which is implementation-independent and whose specifications clearly identify the role of the components and shape their mutual interactions.

Since the "as-a-service" abstraction is considered a key element in the prospected evolution of networking technology, our aim is to take advantage of principles and practices that have emerged in the domain of service lifecycle management in the area of SOA design and implementation.

Moreover, since SOA is an architectural solution which can be mapped into different implementation solutions [41], we deem that the adoption and re-visitation of SOA principles and main architectural guidelines to target our problem domain have the benefit of achieving the desired level of specification abstraction (as mentioned above), while also taking into account architectural patterns and practices developed in more than one decade.

Therefore, our contribution consists first in providing a survey of reference scenarios for virtualized networks and elaborating how SOA, NFV, and SDN provide complementary features toward adaptive composite network service delivery in multi-provider environments. Then, we propose a SOA-inspired architectural model integrating NFV orchestration and SDN network control capabilities, also providing references to related standardization initiatives and illustrating its main workflows. We also describe our ongoing prototyping and testing activities, which aim at validating the proposed approach.

## 3. Reference Service Scenarios

In this section, we survey two main reference service scenarios for virtualized networks made possible by dynamic VNF composition and orchestration (see Figure 1).

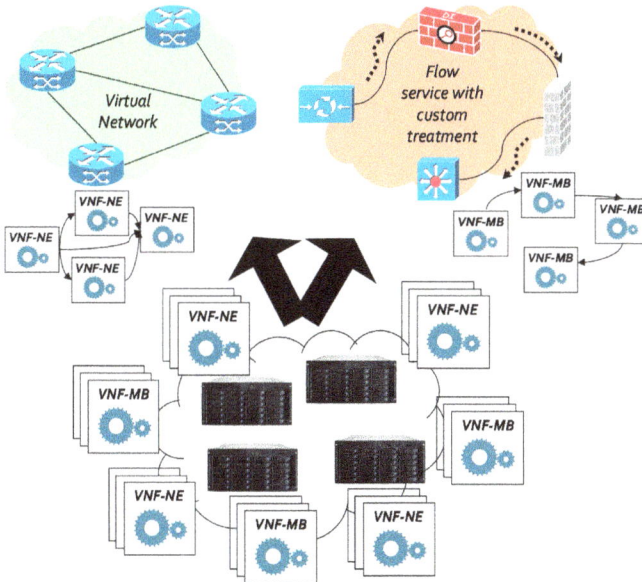

**Figure 1.** Scenarios for dynamic composition of Virtual Network Functions.

The first scenario is enabled by the deployment of VNFs that provide network element (NE) functions, *i.e.*, VNF-NEs. Examples of NEs might be either switching elements (e.g., routers, broadband remote access server), mobile network nodes (HLR/HSS, SGSN, GGSN/PDN-GW, base stations), or signaling control systems (e.g., session border controllers) [12]. Different functions of the same NE can be also partitioned and deployed as different VNFs operated in different network locations instead as one single comprehensive VNF. For instance, a session border controller can be split in the following functions deployed as VNFs, *i.e.*, session terminations executed at the edge of the network, admission control executed in the core network, and statistics and billing data collection executed at a data center [2].

In general terms, the sequence of deployed VNFs formally represents a virtual network (VN) established to deliver different kinds of network infrastructure services (e.g., IMS, 4G, IP/MPLS) and loosely coupled with the exact physical location of constituent VNFs. On the one hand, the deployment of VNF-NE chains allows for NSPs to set-up/upgrade the network infrastructure while taking advantage of the deployment of network functions as virtual machines. On the other hand, the NSPs can enrich their service offerings through the delivery of virtualized network infrastructure services to third-parties, *i.e.*, "VN service". In a scenario where network functions can be dynamically discovered, negotiated and elastically composed as services, application service providers may lease VNF-NE chains with given communication capabilities from different NSPs and compose them to operate an end-to-end virtual service infrastructure to offer value-added application services to users (e.g., delay-optimized infrastructure for high-definition video applications [6]). Accordingly, a request for "VN service" set-up can be internally issued by a Network Management System (NMS) or an Operations Support System (OSS) on behalf of a network planning function of a NSP to establish/update the network infrastructure. Furthermore, a request for a virtual service infrastructure deployment, such as a next generation service overlay network (NGSON), can be issued by an application service provider.

The second scenario derives from the deployment of VNFs providing middlebox (MB) functions, *i.e.*, VNF-MBs. Example of MBs are: firewall, network address translation, WAN optimization controller (WOC), deep packet inspection (DPI), intrusion detection systems, load balancers, multimedia transcoders, virus scanning. Elastic deployments of VNF-MBs allow for differentiated data processing since service data flows can traverse only the needed VNFs, while skipping the unnecessary ones. For instance, in the case of a long-lived multimedia data flow it could be beneficial to avoid a DPI appliance in order to confine delays and save processing resources, while it could be beneficial including a WOC function so that traffic would be routed over links with proper level of delay and jitter guarantees. Accordingly, a service delivery path need to be established to serve an application data flow which includes specified network processing functions to be traversed for addressing given application requirements.

The capability to dynamically establish chains of VNF-MBs allows NSPs enriching their Quality of Service (QoS) offerings through the delivery of paths with extended QoS guarantees that do not only address delay or throughput requirements but also availability, reliability and security requirements demanded by the application data flows, *i.e.*, "flow service with custom data treatment". In fact, in a dynamic landscape of service delivery, VNF-MB chains can be dynamically provisioned that extend the network forwarding capabilities with customizable data processing features. Accordingly, the sequence of deployed VNFs formally represents a request for a "flow service with custom data treatment", which can be issued by a service delivery platform, e.g., NGSON, on behalf of an application service provider during a negotiation phase for adequate transport QoS guarantees [11].

In the rest of the paper, such service scenarios will be generally referred to as network service chaining, irrespective of the fact that VNFs in the chain deploy NE or MB functions.

## 4. Service-Oriented Approach for Dynamic Network Services

The adaptive composition and delivery of network services within the NFV framework requires the conception of architectural models and techniques for dynamic definition, orchestration and management of network functions. We argue that SOA provides an effective solution for coordinating and elastically composing virtualized network functions, *i.e.*, VNFs, across heterogeneous systems while leveraging SDN capabilities to programmatically set up data delivery paths through the dynamically established sequence of VNFs. Thus, as depicted in Figure 2, we envision a synergistic connection among SOA, NFV and SDN for addressing more effective service delivery models that allow NSPs to stay competitive and keep the pace with the service offer and infrastructure innovation of OTT providers.

Although virtualization does not necessarily imply the adoption of service-oriented principles, it favors the usage of service-oriented abstractions to model and handle network functions executed on top of virtual resources. Indeed, thanks to the definition of open and well-defined interfaces between network functions and their management entities, VNFs can be represented as "black boxes" in the form of SOA-compliant network services (arrow 1 in Figure 2). By leveraging service-oriented principles, these network services can be dynamically consumed and composed to provide more complex and adaptive services based on specified requirements, even in a multi-provider environment (arrow 2). To complete the picture, the dynamic provision of VNF-enabled service chains leverages the capability offered by SDN of enforcing adequate traffic forwarding capabilities through the constituent VNFs (arrow 3).

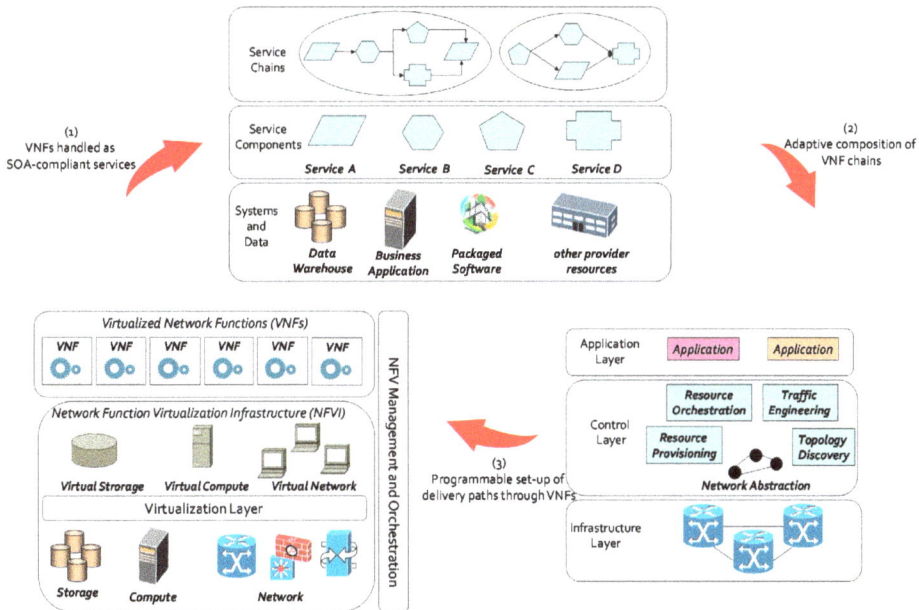

**Figure 2.** Service Oriented Architecture, Network Function Virtualization, and Software Defined Networking synergy for adaptive network service composition and delivery models.

In this perspective, the adoption of service-oriented models can help in defining the proper level of abstraction of network capabilities provided by the NFV infrastructure, thus enabling a loose-coupled, flexible, and effective collaboration among providers of infrastructural resources, network and application services. Indeed, by offering a virtualized access to the physical infrastructure, a NSP can cooperate with other providers to offer advanced network services and capabilities to different consumers (e.g., OTT providers) on top of a virtual resource infrastructure, *i.e.*, NFVI.

*Dynamic Service Chaining of Virtual Network Functions*

The deployment of network functions as VNFs paves the way for network resource and capabilities to be handled as independent and self-contained services that can be flexibly composed to provide a dynamically-established sequence of functionalities leveraging SOA principles, *i.e.*, dynamic service chaining.

In dynamic service chaining, the invocation flow of VNFs is specified at run time for addressing flexible and adaptive service deployments. We define as a "Composite Service" a generalized network service to be provisioned as a result of proper associations of VNF instances according to specified requirements. As depicted in Figure 3, the provisioning of a "Composite Service" through the orchestration of VNFs is enabled through the following functions: "service composition specification and retrieval", "service selection", "service delivery", and "service placement". The first two functions are typical of service-oriented architectures and are here slightly revisited to align with the terminology adopted in the VNF standard specifications. The latter two functions, instead, strongly depend on the peculiarities of reference service scenarios for network service chaining introduced in Section 3.

Service composition specification consists in defining a chaining logic as a workflow of functional network capabilities, *i.e.*, "Abstract Service Chain". More specifically, the "Abstract Service Chain" corresponds to the VNF-graph defined in ETSI [12] and specifies the types of VNFs that should be connected to provide a service (or part of a service), their order and the connectivity requirements among them (e.g., dependencies between VNFs, maximum latency and/or minimum bandwidth of inter-VNF connection, *etc.*), while not specifying yet the location and implementation details of the instances that should actually provide those functions. When a service request has to be handled, the related service composition specification is retrieved. The "Abstract Service Chain" is thus a template for service chain instances. The level of abstraction thus introduced is essential to assure loose coupling between the exposed network service and the specific implementation and management issues of the VNF chain and its constituent elements. This promotes the flexibility to choose among available VNF instances at runtime (*i.e.*, service selection), especially in a multi-provider environment. Furthermore, this level of abstraction eases the delivery of dynamically adaptable services to accommodate policy changes as well as requirements of clients accessing the service. Indeed, context-aware adaptation rules can be applied to modify at run time the "Abstract Service Chain" according to the current situation, e.g., user location and network load [11].

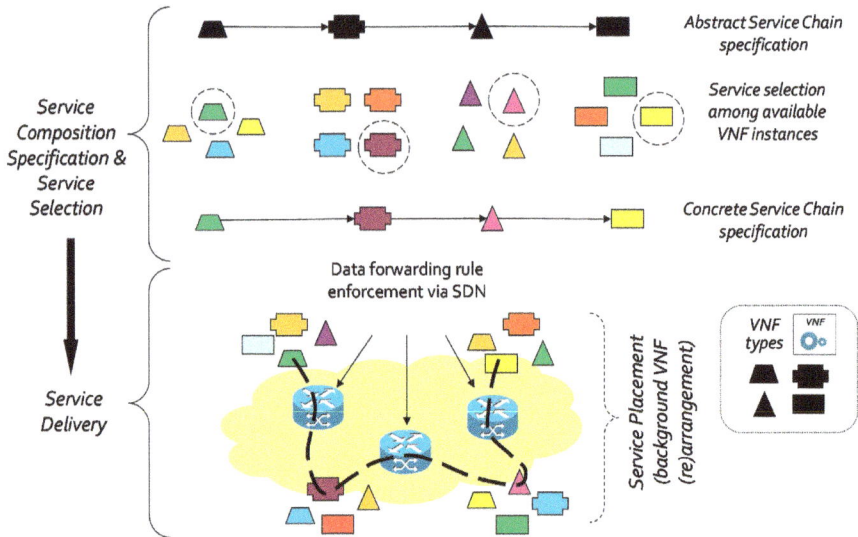

**Figure 3.** Service-oriented mechanisms for dynamic service chaining of Virtual Network Functions (VNFs).

Service selection consists in the proper identification of the sequence of VNF instances among available candidates thereby mapping the "Abstract Service Chain" into a "Concrete Service Chain". Different algorithms can be used to this purpose that optimize a QoS-based utility function (e.g., minimizing the latency per-application traffic flows) for a given composition plan. Such algorithms can consider the computation capabilities and load status of resources executing the

VNF instances, either deduced through estimations from usage historical data or collected through real-time monitoring data (*i.e.*, context-aware selection). For this reason a "Concrete Service Chain" should include references to dynamic information on the status of the service instance and its constituent elements, *i.e.*, monitoring information related to individual VNF instances and links connecting them as well as derived monitoring information at the chain level (e.g., end-to-end delay). At runtime, if one or more VNF instances are no more available or QoS degrades below a given threshold, the service selection task can be rerun to perform service substitution.

Service delivery consists in provisioning delivery paths throughout the selected chain of VNFs thereby accomplishing the required VNF associations. It concerns the use of forwarding functions provided by the underlying network infrastructure to guarantee the proper connectivity among VNF instances specified in the "Concrete Service Chain". The service delivery task is realized through the enforcement of data flow forwarding rules able to address the connectivity requirements among the selected VNF instances. Here SDN plays a key role since it offers the capability to programmatically enforce traffic forwarding rules across network nodes on per-flow or per-tenant basis. Such capability can be exploited to selectively deliver service data through the dynamically established sequence of VNF instances thereby accomplishing the deployment of a network service chain [42,43].

Service Placement consists in the allocation of virtual resources executing VNFs and/or re-arrangement of running VNFs (e.g., migration) while optimizing a specified cost function (e.g., energy consumption, network congestions). Indeed, the VNF instances that take part to a service chain may be deployed in different locations and be provided by different parties. This task runs in the background with respect to the above-mentioned tasks. Indeed, the orchestration assumes that the VNF instances are available at the moment of the service deployment as well as throughout the delivery phase. In order to cope with dynamic application requirements (e.g., user mobility, QoS, or security requirements) or unexpected events (e.g., load congestions, failures), NSPs implement proper resource re-arrangement strategies to deploy VNFs according to the new context while guaranteeing the continuity of the service chain.

## 5. Architectural Design

In this section we present the functional architecture of a NFV orchestrator with SDN network control capabilities and describe the main functional entities (FEs) involved in the proposed VNF orchestration approach.

As shown in Figure 4, we distinguish two main functional network control layers, *i.e.*, the "Network Service Provisioning and Control" and the "Infrastructure Control and Management" layer. The former interacts with an "Application Layer", while the latter interacts with the VNF instances and the underlying resource infrastructure (*i.e.*, the ETSI NFVI).

The "Application Layer" includes instantiation and lifecycle management functions of applications, including network operation support applications. Typically, these functions provide service control and delivery capabilities and are operated at service delivery platforms, e.g., NGSON, or at network service operation systems, e.g., NMS/OSS.

The "Network Service Provisioning and Control" layer supports the dynamic establishment of composite network services on behalf of applications while addressing adequate connectivity requirements. From a SOA perspective, it covers the role of service broker by interfacing with service consumers residing at the "Application Layer", while handling the interactions with VNF instances by leveraging the control and management features provided by the underlying layer.

The "Service Control" Functional Entity (FE) is in charge of coordinating the provision of network services implemented as a chain of VNFs, according to pre-defined "Abstract Service Chain" specifications. It offers a north-bound interface to applications (e.g., NGSON and NMS/OSS) to request the setup of "Composite Services" (e.g., flow with custom data treatment and VN service, respectively). It leverages the "Service Orchestrator" FE capabilities for instantiating a service chain according to the specifications of the selected "Abstract Service Chain".

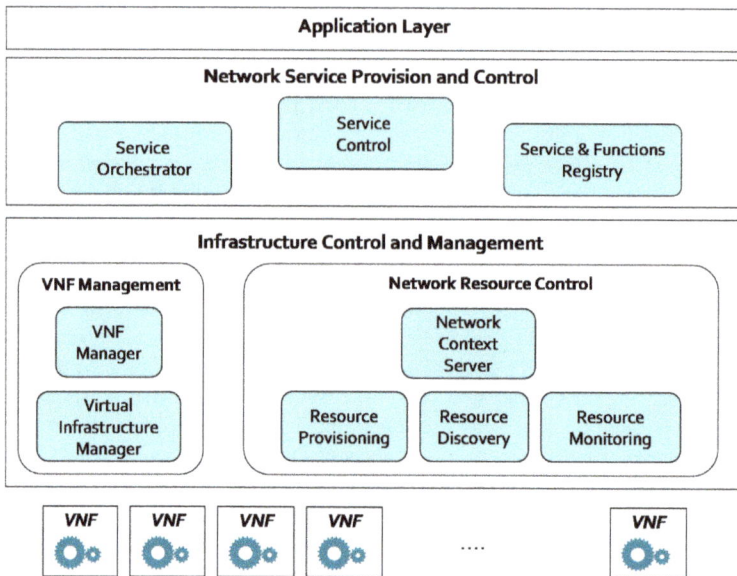

**Figure 4.** Functional architecture for adaptive service composition and delivery.

The "Service Orchestrator" FE maps the "Abstract Service Chain" into a "Concrete Service Chain", thereby selecting the proper VNF instances. Moreover, it coordinates the instantiation, operation and connection of VNFs to satisfy the requirements of instantiated services and to manage adaptation mechanisms for coping with service requirement changes or service degradations. It also instructs the "Network Resource Control" FE for enforcing the proper forwarding rules to establish the delivery path through the specified chain of VNF instances. This process runs transparently to the "Application Layer".

The "Service and Functions Registry" FE maintains the information base required by the "Service Orchestrator" for performing its decision and orchestration tasks. Thus, it handles descriptive and operational information on individual VNFs and on links connecting them (Table 1). Such information is collected through the "VNF Management and Network Context Server" functions of the "Infrastructure Control and Management" layer, described below. If some performance degradation at either VNFs or link is detected, the "Service Orchestrator" is notified. The "Service Orchestrator" performs the appropriate actions for modifying the network service implementation (e.g. substituting VNF instances or the path interconnecting them), otherwise it informs the "Service Control" about the need of cooperative renegotiation with the "Application Layer" functions to reconsider service requirements.

The "Infrastructure Control and Management" functions are responsible for the proper transfer of data across network nodes and VNF management.

The "VNF Management" FE is in charge of managing the lifecycle of individual VNFs and underlying virtual resources. Indeed, it can be decomposed into two main functional components, also according to ETSI MANO specifications: (i) VNF Manager(s), which in accordance with the instructions received by the "Service Orchestrator", initiates or terminate VNF instances or modifies the existing configuration and (ii) virtual infrastructure manager(s), which manages the lifecycle of virtual resources, such as virtual machines and containers. Both components collect monitoring information on the VNFs and resources operational status and performance metrics (Table 1) and notify the "Service and Function Registry" when events of interest occur (e.g., unavailability of a VNF instance).

**Table 1.** Network Function Virtualization data model.

| Entity | Information Category | Description | Data |
|---|---|---|---|
| VNF | descriptive | It contains the information required to automate the instantiation and use of the VNF capability into a service. | VNF id, VNF type, network location and interface capability, requirements for the VNF instantiation and operation (e.g., Virtual Machine specification, required storage and computation resources), management information (e.g., configuration information, initiation and termination scripts), performance indicators, VNF dependencies, horizontal scale thresholds (e.g., maximum load in terms of percentage of nominal computational and traffic capabilities). |
| | operational | It contains the monitoring information of VNFs in a running service instance. | Operational status (e.g., start, stop, pause), usage of CPU, memory and storage resources, traffic load at network interfaces. |
| VNF link | descriptive | It contains the information related to the connectivity (established or to be established) between two or more VNFs, possibly including NFs. | Link type (e.g., point-to-point *vs.* point-to-multipoint, logical link among VNFs *vs.* physical link between a VNF and a Network Function), performance indicators (e.g., QoS parameters). |
| | operational | It contains the monitoring information of established links between VNFs in a running service instance. | Delay, throughput. |

The "Network Resource Control" FE is in charge of setting up and managing the network links required to connect the VNF instances. It handles the requests originated from the "Service Orchestrator" FE for connecting a set of (newly) instantiated VNFs for new (or modified) service contracts by properly enforcing the forwarding rules in the network devices along the delivery path. More specifically, the "Resource Discovery" FE handles the discovery of resource capabilities and network topology thus enabling routing, path computation and traffic engineering decisions. The "Resource Provisioning" FE handles low-level configuration directives to addresses the programming of the network nodes based on information provided by the "Resource Discovery" FE and under the coordination of the "Service Control" FE to consistently steer data flows through the desired VNF chain. The "Resource Monitoring" FE handles low-level directives for network monitoring (*i.e.*, collection of traffic statistics) to feed the "Network Context Server" FE, which is in charge of collecting the network topological and dynamic information of VNF links and making it available to the "Services and Functions Registry" FE with the proper level of abstraction and granularity. More specifically, it handles the descriptive and operational information of links connecting VNFs (Table 1). Efforts for defining VNF and VNF link data models are on-going both in ETSI [12] and in IETF [44] standardization bodies.

The "Network Resource Control" FE and the "Network Context Server" FE can be identified within the IETF ABNO architecture. Specifically, the "Network Context Server" FE refers to the IETF Application-Layer Traffic Optimization (ALTO) server [45] while the "Network Resource Control" FE mainly refers to the ABNO controller. The ALTO server is envisioned to provide abstract network topology information on an end-to-end basis (e.g., network location structure, topological distance between locations and cost between them) for allowing applications to take decisions about service deployments. Relevant network costs information is, for instance, the maximum bandwidth, minimum cross-domain traffic, and lower cost to the user.

The ABNO controller is the main component of the ABNO architecture [13] and is responsible for orchestrating the workflows among ABNO components (e.g., path computation element, topology managers policy agents, and provisioning manager) while addressing application requests of end-to-end network services, e.g., point-to-point path between specified endpoints. Moreover, the ABNO controller is envisioned to provide dynamic network performance information to applications, such as bandwidth usage, available capacity and end-to-end delay.

In Figure 5 the sequence diagram of orchestration workflows is reported for the considered reference service scenarios. The dotted boxes depict the workflows related to the background collection of monitoring information (top) and to the service recovery in case of degradation (bottom), while the main workflow shows the main interactions among FEs for processing a VNF chain set-up request as a result of the orchestration process. More specifically, the main workflow is triggered by a service request issued by an application client to the "Service Control" FE (arrow 1). This FE first retrieves the abstract service description that matches the received request and asks the "Service Orchestrator" FE to return a concrete service chain for that abstract service (arrow 2). The service orchestrator selects the VNF instances that implement the concrete service chaining by properly taking into account current context information, such as VNF and network monitoring data (arrows 3 and 4). Then, it asks the "Network Resource Control" FE for setting up the data path connecting the VNF instances in the service chain (arrows 5–6) and, finally, returns a response to the "Service Control" FE (arrow 7). The workflow in the upper dotted box shows how the Service and Functions registry receives updates on the status of VNF and infrastructure resources from the "VNF Management" FE (arrow 1.a) and on the status of the network (arrows 2.a and 3.a). The workflow in the lower dotted box shows an example of service chain dynamic adaptation to possible network service performance metrics degradation. In the depicted example, we suppose that the service degradation is due to the network path status (e.g., congestion at switches) and, therefore, the "Service Orchestrator" FE, upon the notification issued by the "Service and Functions Registry" (arrow 1.b) requests the setup of a new path connecting the VNF instances (arrows 2.b and 3.b). Of course, more complex workflow could be executed for the service chain dynamic adaptation (e.g., for substituting a subset of VNF instances).

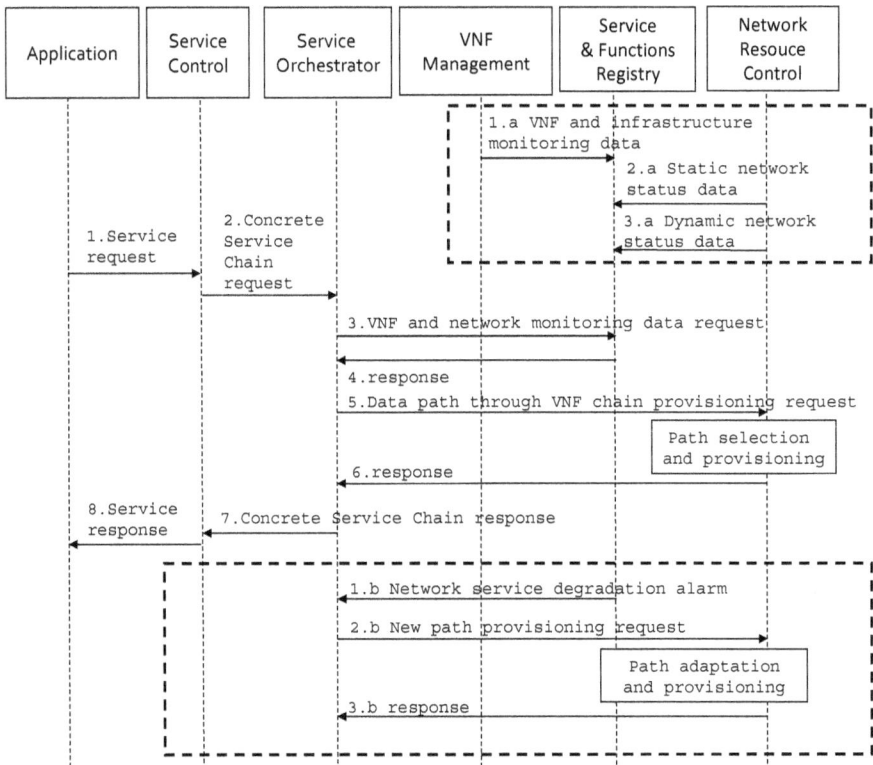

**Figure 5.** Orchestration workflows for adaptive service delivery.

## 6. Architecture Preliminary Validation via Prototype

In order to provide a preliminary validation of the proposed approach we implemented a prototype that provides a subset of the functionalities of the service-oriented architecture presented above. The prototype consists of an SDN controller enhanced with orchestration capabilities implemented on top of a Floodlight controller [46] to establish dynamic chains of network functions (e.g., middlebox). Basically, the prototype plays the role of an orchestrator for dynamic service chaining operating over SDN networks, *i.e.*, SDN orchestrator. In particular, the SDN orchestrator addresses the second service scenario envisioned in the Section 3, *i.e.*, "flow service with custom data treatment", and performs functions both at the "Network Service Provision and Control" layer and at the "Infrastructure Control and Management" layer.

### 6.1. SDN Orchestrator

The software design of the SDN orchestrator is shown in Figure 6b. More specifically, a subset of the "Service Orchestrator" functions is implemented in the request manager, the service data delivery control and coordination and the adaptation module. In fact these blocks handle requests for the establishment of data delivery paths throughout a specified set of middlebox services. These blocks leverage the basic network server (BNServer) and the Floodlight components that implement the "Network Resource Control" functions. As a result of interactions between the above blocks triggered by path set-up requests, a consistent set of forwarding rules can be enforced across network nodes in order to steer service data flows through the desired VNF chain. In addition to the ordinary workflow for the setup of a new request, the data delivery path set-up requests can be also generated by other SDN orchestrators for the establishment of paths to be concatenated to other paths in a multi-provider scenario or by other internal blocks as a result of an orchestration process to adapt established paths to modified network status. The forwarding rules are enforced in the network nodes via the south-bound interface (SB-I) through the OpenFlow protocol [47]. Moreover, these blocks put in place adaptation mechanisms to cope with degradation events. In fact, degradation events are likely to occur as a result of concurrent usage of network node capabilities among multiple services and the prototype allows for reactively adapting loads across network nodes while addressing required data delivery performance. To this purpose, the prototype also includes "Network Context Server" functions implemented by the statistics collector block, which collects, elaborates, and aggregates OpenFlow statistics about network nodes and links (e.g., switch throughput computed from per-port byte counters), while making them available with the proper level of granularity in internal databases acting as "Service and Function Registry". More specifically, such databases contain descriptive information (e.g., IP addresses of end-points, identifier and port number of intermediate network nodes) as well as operational data about the data delivery (segment) paths established throughout the network realizing (part of) running service chain instances, e.g., delay on a per-chain basis. Such data are elaborated by the adaptation module that is in charge of triggering the service data delivery control and coordination block for the adaptation of (part of) service paths throughout different set of switches (*i.e.*, redirection) as soon as a degradation event is detected (e.g., switch load exceeding a certain threshold).

Without lack of generality, the only orchestration and adaptability at the level of the data service delivery path is addressed by the SDN orchestrator prototype. The adaptability also at the level of VNFs and the respective lifecycle management via a cloud management platform is considered as future work. Further details on the prototype design, functional blocks and implementation can be found in [48].

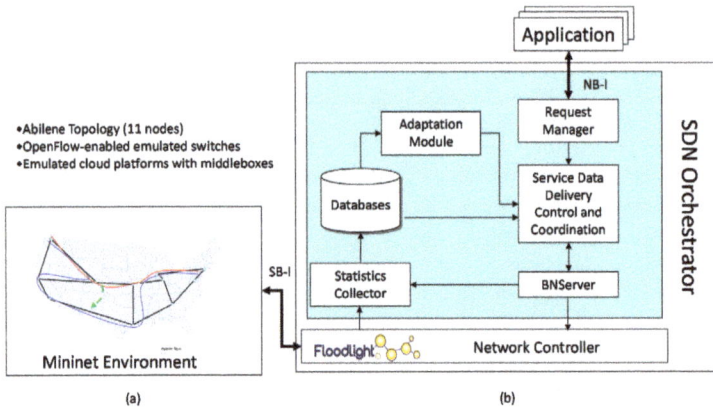

**Figure 6.** Prototype design and emulated testbed environment. (**a**) Emulation environment adopting the Mininet tool and the Abilene network topology; (**b**) architecture of the SDN Orchestrator prototype.

## 6.2. Experimental Results

In this subsection we describe the tests that we carried out to verify, first, the operation of the prototype and, then, the advantages and disadvantages of the orchestration approach, as implemented in the prototype. Figure 6a shows the adopted emulation environment, which uses Mininet, a network emulator for deploying large networks [49]. In line with other current works in literature [43], we have used the Abilene topology which is a high-performance backbone network created by the Internet2 community [50] as a reference network. Thus, 11 switches have been deployed and collocated at as many nodes as the Abilene topology. A subset of the switches are connected to emulated cloud platforms that contain the middleboxes, *i.e.*, VNF cloud, whether the remaining switches are simply transit switches. The emulated cloud platforms are supposed to contain the same types of middlebox services. For each middlebox service instance (*i.e.*, VNF) to select, we exploit the Dijkstra's shortest-path algorithm provided by the Floodlight controller to find the network path connected to the closest switch associated with a cloud platform. Each instance of middlebox, *i.e.*, VNF, specified in the request is selected from different cloud platforms. Moreover, all the nodes in the topology are randomly chosen to behave as a source/destination of the service delivery path requests.

In the first round of experiments we used 5 bursts of 100 requests each to collect results on the general performance of the SDN orchestrator. We fixed the number of switches connected to a cloud platform to five (*i.e.*, five out of 11 switches connected to a VNF cloud), we vary the number of middlebox services in the request (*i.e.*, VNFs in the chain) and we measured the average number of flow entries in the switches and the network set-up time (*i.e.*, time required to setup the flow entries in all the switches across the chosen path). From Table 2, we notice that both the number of flow entries installed in these switches (truncated value) and the network set-up time increases with the number of required services in a chain due to the higher number of traversed switches (from two to five) and, thus, to the higher number of flow entries to be setup to forward the traffic along the data delivery path. Then, we fixed the number of requested virtual functions in a chain to three and we vary the number of VNF clouds in the network. From Table 3, we notice that increasing the number of available cloud platforms and spreading them on different switches decrease the number of paths traversing the same switch. This alleviates the load (*i.e.*, in terms of number of flow entries) on the switches and prevents from possible failures or packets loss. Additionally, the network setup time decreases as the more VNF clouds are available in the network while showing to be less dependent on respect the number of VNF clouds. In fact, in this case the total path length differs just due to the decreasing need to traverse additional transit switches because the number of VNFs in the chain is the same.

**Table 2.** Orchestrator performance with different number of VNFs in the chain and with five VNF clouds.

| Number of VNFs in the Chain | Average Number of Flow Entries | Network Set-up Time |
|:---:|:---:|:---:|
| 3 | 153 | 36.32 ms |
| 5 | 253 | 50.14 ms |

**Table 3.** Orchestrator performance with different number of VNF clouds and with three VNFs in the chain.

| Number of VNF Clouds in the Network | Average Number of Flow Entries | Network Set-up Time |
|:---:|:---:|:---:|
| 3 | 249 | 43.64 ms |
| 6 | 132 | 36.05 ms |

In the second round of experiments we evaluated the effectiveness of the orchestration feature in terms of level of usage of switches and in terms of overhead due to data delivery path redirections. To this purpose, we carried out five tests using sequences of 100 requests. Each request is characterized by a random source, a random destination and 4 VNFs. The number of switches connected to a VNF cloud is equal to five. The request follows a Poisson distribution characterized by an average inter-arrival time of 20 s and an average flow duration of 20 s.

Table 4 shows the average number of transmitted bytes along with the standard deviation related to the switches co-located with the VNF clouds. We can observe that the redirection of path has beneficial effects on the overall load of switches since the number of transmitted bytes is more fairly-distributed among all the available switches. This is in line with the IETF guidelines for the control functionalities governing the service function chaining [51]. It is worth highlighting that while addressing data delivery performance, our approach preserves the perceived quality of the services since no packets are lost during the tear-down of the data delivery paths that are rapidly re-established through unloaded switches. However, this enhanced performance is obtained at the cost of the increased number of exchanged messages and the presence of redirection time, as shown in Table 5 in comparison with a case without redirection considered in the literature [43].

**Table 4.** Switch load with and without path redirection.

| Redirection Option | Average Number of Transmitted Bytes | Standard Deviation |
|:---:|:---:|:---:|
| With redirection feature | 23 MBytes | 5.56 |
| Without redirection feature | 43 MBytes | 34.43 |

The redirection feature increases the number of messages exchanged internally since every time a redirection is performed, a new path is calculated and new flow entries are setup which necessitates further communication messages (e.g., path delivery set-up, path teardown, *etc.*). Moreover, these operations require a certain time, evaluated to 0.3 s in these tests, which is acceptable with respect to the overall flow duration.

**Table 5.** Redirection overhead.

| Redirection Option | Number of Exchanged Messages | Redirection Time |
|:---:|:---:|:---:|
| With redirection feature | 11,195 | 30 ms |
| Without redirection feature | 3449 | 0 |

## 7. Conclusions

In this paper we discussed how service-oriented principles can be applied to effectively orchestrate virtualized network resources, *i.e.*, VNFs, to provide dynamically-established VNF chains while taking full advantage of programmable data forwarding capabilities provided by SDN to adaptively deliver data throughout VNFs. The level of abstraction introduced by SOA principles provides a generalized mechanism for composing heterogeneous resources (in the computing, as well as in the networking domain) across different providers to provide users with an enhanced service experience. Moreover, the exposure of VNFs as independent services advertised through a service contract description promotes the delivery of complex network services implemented as a composition of dynamically-selected and bound VNFs. In addition, composition adaptation and service selection and substitution mechanisms can be put in place to assure the provision of QoS-aware dynamic adaptive services. We also present current results of our ongoing activities in the implementation of a prototype to validate the proposed approach. In the near future we plan to extend the prototype with VNF and virtual infrastructure management capabilities.

This approach can give several benefits to NSPs: a faster innovation speed of their network infrastructure without the need of radically changing hardware systems, reduction in the time and the cost for rolling out new services, adaptive network service provisioning and, finally, new business opportunities fostered by a dynamic environment of multi-provider cooperation.

The proposed approach poses a number of challenges for NFV and SDN research areas. Firstly, efficient mechanisms for the automated orchestration of VNF instances and their placement are needed for adaptively addressing service requirements. Indeed, optimization algorithms would be required to provide the best set of VNF instances (in case of orchestration) and the set of locations where to put VNFs (in case of placement) while minimizing a cost function, e.g., overall latency along the VNF path. Secondly, in the SDN area the main envisioned challenge is how to provide dynamic and granular traffic steering capabilities while scaling at the tenant or application flow level. In both areas, high-scale monitoring functions are required for tracing the actual service availability and properly trigger service recovery operations. Moreover, such data need to be exposed to the orchestration functions with the proper level of abstraction and granularity to keep the problem scalable [2].

**Acknowledgments:** We acknowledge Molka Gharbaoui and Ahmed Mohammed Ali for their technical assistance.

**Author Contributions:** The authors equally contributed to this work.

**Conflicts of Interest:** The authors declare no conflict of interest.

## Abbreviations

The following abbreviations are used in this manuscript:

| | |
|---|---|
| ABNO | Application-Based Network Operations |
| DPI | Deep Packet Inspection |
| ETSI | European Telecommunication Standard Institute |
| FE | Functional Entity |
| GGSN | Gateway GPRS (General Packet Radio Service) Service Node |
| HLR | Home Location Register |
| HSS | Home Subscriber Server |
| IETF | Internet Engineering Task Force |
| IMS | IP Multimedia Subsystem |
| IP/MPLS | Internet Protocol/Multi-Protocol Label Switching |
| MB | middlebox |
| NE | Network Element |
| NFV | Network Function Virtualization |
| NGSON | Next Generation Service Overlay Network |

| NSP | Network Service Providers |
|-----|---------------------------|
| PDN-GW | Packet Data Network Gateway |
| SDN | Software-Defined Network |
| SGSN | Serving GPRS (General Packet Radio Service) Service Node (aka Serving GPRS Support Node) |
| SOA | Service Oriented Architecture |
| OTT | Over-The-Top |
| VN | Virtual Network |
| VNF | Virtual Network Function |
| WOC | WAN optimization controller |
| 4G | 4 Generation |

## References

1. Han, B.; Gopalakrishnan, V.; Ji, L.; Lee, S. Network function virtualization: Challenges and opportunities for innovations. *IEEE Commun. Mag.* **2015**, *53*, 90–97. [CrossRef]
2. GEx Multi-Domain Service Creation—From 90 Days to 90 Minutes. White Paper. March 2016. Available online: http://www.5gex.eu/wp/wp-content/uploads/2016/03/5GEx-White-Paper-v1.pdf (accessed on 20 April 2016).
3. John, W.; Pentikousis, K.; Agapiou, G.; Jacob, E.; Kind, M.; Manzalini, A.; Meirosu, C. Research directions in Network Service Chaining. In Proceedings of the 2013 IEEE SDN for Future Networks and Services (SDN4FNS), Trento, Italy, 11–13 November 2013; pp. 1–7.
4. Nunes, B.; Mendonca, M.; Nguyen, X.; Obraczka, K.; Turletti, T. A survey of Software-Defined Networking: Past, present, and future of Programmable Networks. *IEEE Commun. Surv. Tutor.* **2014**, *16*, 1617–1634. [CrossRef]
5. Europen Telecommunications Standard Institute (ETSI). *Network Functions Virtualisation (NFV); Architectural Framework*; GS NFV 002 V1.2.1; ETSI: Sophia Antipolis Cedex, France, 2014.
6. Latre, S.; Famaey, J.; de Turck, F.; Demeester, P. The fluid internet: Service-centric management of a virtualized future internet. *IEEE Commun. Mag.* **2014**, *52*, 140–148. [CrossRef]
7. Duan, Q.; Yan, Y.; Vasilakos, A.V. A Survey on Service-Oriented Network Virtualization toward convergence of networking and cloud computing. *IEEE Trans. Netw. Serv. Manag.* **2012**, *9*, 373–392. [CrossRef]
8. Erl, T. *SOA, Principles of Service Design*; Prentice Hall: Upper Saddle River, NJ, USA, 2008.
9. Branca, G.; Atzori, L. A survey of SOA technologies in NGN Network Architectures. *IEEE Commun. Surv. Tutor.* **2012**, *14*, 644–661. [CrossRef]
10. Pentikousis, K.; Meirosu, C.; Lopez, D.R.; Denazis, S.; Shiomoto, K.; Westphal, F.J. Guest editorial: Network and service virtualization. *IEEE Commun. Mag.* **2015**, *53*, 88–89. [CrossRef]
11. Paganelli, F.; Ulema, M.; Martini, B. Context-aware service composition and delivery in NGSONs over SDN. *IEEE Commun. Mag.* **2014**, *52*, 97–105. [CrossRef]
12. ETSI. *Network Functions Virtualisation (NFV); Management and Orchestration*; GS NFV-MAN 001 V1.1.1; ETSI: Sophia Antipolis Cedex, France, 2014.
13. ETSI. *Network Functions Virtualisation (NFV); Infrastructure; Network Domain*; GS NFV-INF 005 V1.1.1; ETSI: Sophia Antipolis Cedex, France, 2014.
14. ETSI. *Network Functions Virtualisation (NFV); Ecosystem*; Report on SDN Usage in NFV Architectural Framework; GS NFV-EVE 005 V1.1.1; ETSI: Sophia Antipolis Cedex, France, December 2015.
15. Quinn, P.; Nadeau, T. Problem Statement for Service Function Chaining. RFC 7498. April 2015. Available online: https://tools.ietf.org/htm/rfc7498 (accessed on 20 April 2016).
16. Network Function Virtualization Research Group. Available online: https://irtf.org/nfvrg (accessed on 6 May 2016).
17. King, D.; Farrel, A. A PCE-Based Architecture for Application-Based Network Operations. Available online: http://tools.ietf.org/html/rfc7491 (accessed on 31 May 2016).
18. IEEE Std. 1903–2011, Standard for the Functional Architecture of Next Generation Service Overlay Networks. Available online: https://standards.ieee.org/findstds/standard/1903-2011.html (accessed on 6 May 2016).
19. IEEE Software Defined Networks (SDN). Available online: http://sdn.ieee.org/ (accessed on 6 May 2016).

20. IEEE. Working Toward the Next Generation of Networks. Available online: http://theinstitute.ieee.org/benefits/standards/working-toward-the-next-generation-of-networks (accessed on 6 May 2016).
21. Open Networking Foundation. SDN Architecture—Issue 1 TR-502, June 2014. Available online: https://www.opennetworking.org/images/stories/downloads/sdn-resources/technical-reports/TR_SDN_ARCH_1.0_06062014.pdf (accessed on 20 April 2016).
22. Open Networking Foundation. SDN Architecture—Issue 1.1 TR-521, January 2016. Available online: https://www.opennetworking.org/images/stories/downloads/sdn-resources/technical-reports/TR-521_SDN_Architecture_issue_1.1.pdf (accessed on 20 April 2016).
23. Open Networking Foundation. Available online: https://www.opennetworking.org/technical-communities/areas/services/2860-cross-stratum-orchestration-cso (accessed on 6 May 2016).
24. Open Networking Foundation. OpenFlow-Enabled SDN and Network Functions Virtualization. Available online: https://www.opennetworking.org/images/stories/downloads/sdn-resources/solution-briefs/sb-sdn-nvf-solution.pdf (accessed on 20 April 2016).
25. Broadband Forum. Available online: https://www.broadband-forum.org/ (accessed on 6 May 2016).
26. Framework 15.0 Foundational Studies. Available online: https://www.tmforum.org/zoom/frameworx-15-0-foundational-studies (accessed on 6 May 2016).
27. Rosa, R.V.; Silva Santos, M.A.; Rothenberg, C.E. MD2-NFV: The case for multi-domain distributed network functions virtualization. In Proceedings of the IEEE 2015 International Conference and Workshops on Networked Systems (NetSys), Cottbus, Germany, 9–12 March 2015; pp. 1–5.
28. Lopez, V.; de Dios, O.G.; Fuentes, B.; Yannuzzi, M.; Fernández-Palacios, J.P.; Lopez, D. Towards a network operating system. In Proceedings of the 2014 Optical Fiber Communication Conference, San Francisco, CA, USA, 9–13 March 2014; pp. 1–3.
29. Naudts, B.; Tavernier, W.; Verbrugge, S.; Colle, D.; Pickavet, M. Deploying SDN and NFV at the speed of innovation: Toward a new bond between standards development organizations, industry fora, and open-source software projects. *IEEE Commun. Mag.* **2016**, *54*, 46–53. [CrossRef]
30. Giannoulakis, I.; Kafetzakis, E.; Xylouris, G.; Gardikis, G.; Kourtis, A. On the applications of efficient NFV management towards 5G networking. In Proceedings of the 1st International Conference on 5G for Ubiquitous Connectivity (5GU), Akaslompolo, Finland, 26–28 November 2014; pp. 1–5.
31. Sonkoly, B.; Szabo, R.; Jocha, D.; Czentye, J.; Kind, M.; Westphal, F.J. UNIFYing cloud and carrier network resources: An architectural view. In Proceedings of the 2015 IEEE Global Communications Conference (GLOBECOM), San Diego, CA, USA, 6–10 December 2015; pp. 1–7.
32. Giotis, K.; Kryftis, Y.; Maglaris, V. Policy-based orchestration of NFV services in Software-Defined Networks. In Proceedings of the 2015 1st IEEE Conference on Network Softwarization (NetSoft), London, UK, 13–17 April 2015; pp. 1–5.
33. Muñoz, R.; Vilalta, R.; Casellas, R.; Martinez, R.; Szyrkowiec, T.; Autenrieth, A.; López, D. Integrated SDN/NFV management and orchestration architecture for dynamic deployment of virtual SDN control instances for virtual tenant networks [invited]. *IEEE/OSA J. Opt. Commun. Netw.* **2015**, *7*, B62–B70. [CrossRef]
34. Soares, J.; Goncalves, C.; Parreira, B.; Tavares, P.; Carapinha, J.; Barraca, J.P.; Sargento, S. Toward a TELCO cloud environment for service functions. *IEEE Commun. Mag.* **2015**, *53*, 98–106. [CrossRef]
35. Garay, J.; Matias, J.; Unzilla, J.; Jacob, E. Service description in the NFV revolution: Trends challenges and a way forward. *IEEE Commun. Mag.* **2016**, *54*, 68–74. [CrossRef]
36. Mehraghdam, S.; Keller, M.; Karl, H. Specifying and placing chains of virtual network functions. In Proceedings of the 2014 IEEE 3rd International Conference on Cloud Networking (CloudNet), Luxembourg, 8–10 October 2014; pp. 7–13.
37. Moens, H.; de Turck, F. VNF-P: A model for efficient placement of virtualized network functions. In Proceedings of the 2014 10th International Conference on Network and Service Management (CNSM), Rio de Janeiro, Brazil, 17–21 November 2014; pp. 418–423.
38. Abujoda, A.; Papadimitriou, P. MIDAS: Middlebox discovery and selection for on-path flow processing. In Proceedings of the 7th IEEE International Conference on Communication Systems and Networks (COMSNETS 2015), Bangalore, India, 6–10 January 2015.
39. Ferrer Riera, J.; Hesselbach, X.; Escalona, E.; Garcia-Espin, J.A.; Grasa, E. On the complex scheduling formulation of virtual network functions over optical networks. In Proceedings of the 2014 16th International Conference on Transparent Optical Networks (ICTON), Graz, Austria, 6–10 July 2014; pp. 1–5.

40. Lombardo, A.; Manzalini, A.; Riccobene, V.; Schembra, G. An analytical tool for performance evaluation of software defined networking services. In Proceedings of the 2014 IEEE Network Operations and Management Symposium (NOMS), Krakow, Poland, 5–9 May 2014; pp. 1–7.
41. Stal, M. Using architectural patterns and blueprints for service-oriented architecture. *IEEE Softw.* **2006**, *23*, 54–61. [CrossRef]
42. Matias, J.; Garay, J.; Toledo, N.; Unzilla, J.; Jacob, E. Toward an SDN-enabled NFV architecture. *IEEE Commun. Mag.* **2015**, *53*, 187–193. [CrossRef]
43. Zhang, Y.; Beheshti, N.; Beliveau, L.; Lefebvre, G.; Manghirmalani, R.; Mishra, R.; Patney, R.; Shirazipour, M.; Subrahamaniam, R.; Truchan, C.; *et al.* StEERING: A software-defined networking for inline service chaining. In Proceedings of the 2013 21st IEEE International Conference on Network Protocols (ICNP), Goettingen, Germany, 7–10 October 2013.
44. Xu, W.; Jiang, Y.; Zhou, C. Data Models for Network Functions Virtualization. Available online: https://datatracker.ietf.org/doc/draft-xjz-nfv-model-datamodel/ (accessed on 31 May 2016).
45. Seedorf, J.; Burger, E. Application-Layer Traffic Optimization (ALTO) Problem Statement. Available online: https://tools.ietf.org/html/rfc5693 (accessed on 31 May 2016).
46. Floodlight Openflow Controller. Available online: http://www.projectfloodlight.org/floodlight/ (accessed on 6 May 2016).
47. Openflow Specifications. Available online: http://www.openflow.org (accessed on 6 May 2016).
48. Mohammed, A.A.; Gharbaoui, M.; Martini, B.; Paganelli, F.; Castoldi, P. SDN controller for Network-aware Adaptive Orchestration in Dynamic Service Chaining. In Proceedings of the NetSoft 2016, Seoul, Korea, 6–10 June 2016. (to appear).
49. Mininet. Available online: http://www.mininet.org (accessed on 6 May 2016).
50. Abilene Network. Available online: https://en.wikipedia.org/wiki/AbileneNetwork (accessed on 6 May 2016).
51. Boucadair, M., Ed.; Service Function Chaining (SFC) Control Plane Components & Requirements. Available online: https://tools.ietf.org/html/draft-ww-sfc-control-plane-04 (accessed on 6 May 2016).

*future internet*

MDPI

*Review*

# Cognitive Spectrum Sharing: An Enabling Wireless Communication Technology for a Wide Use of Smart Systems

Romano Fantacci *,† and Dania Marabissi †

Department of Information Engineering, University of Florence, Florence 50139, Italy; dania.marabissi@unifi.it
* Correspondence: romano.fantacci@unifi.it; Tel.: +32-055-2758621
† These authors contributed equally to this work.

Academic Editor: Dino Giuli
Received: 2 February 2016; Accepted: 11 May 2016; Published: 20 May 2016

**Abstract:** A smart city is an environment where a pervasive, multi-service network is employed to provide citizens improved living conditions as well as better public safety and security. Advanced communication technologies are essential to achieve this goal. In particular, an efficient and reliable communication network plays a crucial role in providing continue, ubiquitous, and reliable interconnections among users, smart devices, and applications. As a consequence, wireless networking appears as the principal enabling communication technology despite the necessity to face severe challenges to satisfy the needs arising from a smart environment, such as explosive data volume, heterogeneous data traffic, and support of quality of service constraints. An interesting approach for meeting the growing data demand due to smart city applications is to adopt suitable methodologies to improve the usage of all potential spectrum resources. Towards this goal, a very promising solution is represented by the Cognitive Radio technology that enables context-aware capability in order to pursue an efficient use of the available communication resources according to the surrounding environment conditions. In this paper we provide a review of the characteristics, challenges, and solutions of a smart city communication architecture, based on the Cognitive Radio technology, by focusing on two new network paradigms—namely, Heterogeneous Network and Machines-to-Machines communications—that are of special interest to efficiently support smart city applications and services.

**Keywords:** smart communications; cognitive radio; Heterogeneous Networks; machine-to-machine communications; spectrum sharing; spectrum sensing

**PACS:** J0101

## 1. Introduction

The worldwide urbanization process represents a formidable challenge and attracts attention toward cities. In particular, in the near future, the quality of life of billions of people will depend on capability of cities of saving energy, reducing harmful emissions, improving the living conditions and increasing citizens security. These challenges need to be addressed through the implementation of Information and Communication Technology (ICT) intelligent solutions in the urban ecosystem. Indeed, the smart city concept is based on functional integration of software systems, network infrastructures, heterogeneous user devices, and collaboration technologies. Motivated by the fact that a reliable, robust, secure, and scalable communication architecture plays a crucial role for the successful operation of a smart environment, this paper provides an overview of the communication infrastructure of a smart city and the related enabling technologies. First, a review of main characteristics, challenges, and proposed solutions is provided. Then, the focus of the paper is moved on cognitive radio (CR) technology, which is considered an efficient methodology to address communication needs and challenges in many smart city contexts. In particular, in the final part of the paper, CR technology is investigated as a viable solution for mitigating the spectrum shortage that can arise from the use of two key technologies—Heterogeneous Networks (HetNets) and Machine-to-Machine (M2M) communications—that nowadays are envisaged as two promising communication paradigms for future smart cities, specifically characteristics, challenges, and literature solutions are discussed.

## 2. The Communication Infrastructure of a Smart City

The communication infrastructure of a smart city must be suitably designed to satisfy the specific requirements and needs of the considered environment. In particular, it has to support basic functionalities such as sensing, transmission, and control. Sensing is carried out by a large number of sensors and smart devices (even people can be sensors) that monitor the environment; this information is transferred to a control center that performs data elaboration, thus providing control instructions that are delivered to sensors/actuators. A basic characteristic of a smart city is that it is composed of multiple heterogeneous devices that are connected regardless of their locations in an autonomous and scalable way compliant with the Internet of Things (IoT) paradigm. Moreover, a smart environment usually spreads over large geographical areas, and hence requires a multilayer communication infrastructure that extends across the whole smart environment. At least three communication layers can be foreseen: "local area networks", "access networks", and "computing/application networks". The local area networks layer is responsible for monitoring the environment using heterogeneous devices and sensors (which can be embedded in vehicles, buildings, urban environment, and people, for example).The access layer represents the communication backbone that allows the transfer of information from the local networks to the control center. The computing layer is related to the applications that address the data collected by the sensing layer. Even if in some cases wired connections are possible, mobile and ubiquitous wireless connections are preferred for a wide class of services and applications. Mobile connections have the potential to provide remote control and monitoring without the addition of any cabling cost.

The previously described multilayer-structure can provide efficient, reliable, and secure communications if it is designed taking into account the specific needs and challenges of a smart city. Moreover, as a final consideration, we have to note that due to the complexity of the network infrastructure and its deployment costs, a suitable planning in relation to the short, medium, and long-term demands of smart services and applications must be pursued. In particular, future cellular networks should enable to realize a truly networked society with unlimited access to information for anyone. In that way, it will be possible to support various smart infrastructures and smart cities that are green, safe, mobile, connected, and informed.

### 2.1. Communication Challenges

The main challenges the smart city communication infrastructure has to deal with can be summarized as:

- Heterogeneity: heterogeneous communication technologies have to be integrated to provide reliable and functional access to the system elements in different environments. Indeed, in order to have a wide spread of smart city applications, it is required that people can use their own devices and not dedicated devices and software. This will allow low cost, flexible, and scalable solutions. The resulting communication infrastructure must integrate any technology that may be considered relevant by a smart city actor.
- Quality of Service: the diversity of smart city solutions determines the wide variability of supported services and applications. This leads to multiple traffic types within the network and, hence, the need to manage all of these respecting their different Quality of Service (QoS) requirements in terms of priority, delay, data rate, reliability, and security. Moreover, the amount

of data varies tremendously during a day, so the traffic conditions change quickly, and the system must be able to adapt itself to the scenario variability.

- Security: smart city networks have to carry reliable and real-time information toward monitoring and control centres. This exposes the system to outside attacks, unauthorized accesses, and data modifications. Data can be captured and carried over the system. This makes it necessary to foresee suitable mechanisms to prevent cyber attacks that might block the city functionalities and carry unwanted alarms or data theft.

- Energy consumption: smart cities are based on large diffusion of smart devices and sensors whose operations are strongly affected by their battery life. Hence, energy efficient communication protocols are needed, especially for local connections. Moreover, the use of energy harvesting solutions should be considered.

- Communication resource availability: smart cities and smart devices will have an explosive growth in the next years. Hence, the traffic generated by the applications running on smart systems will require a huge amount of bandwidth and network resources, thus challenging the communication infrastructure. Especially in wireless communications, having sufficient and dedicated spectrum resources for smart city applications will be infeasible. Therefore, the availability of sufficient spectrum to accommodate current and future needs of smart environments is expected to be a critical requirement.

*2.2. Related Works*

The problem of communication infrastructures able to efficiently support smart city applications and services in complex, distributed, and diverse environments has been, and continues to be, the subject of intense research investigations. For this reason, in the literature there are several papers related to this topic, especially for the smart grid context. Among others, the aim of [1,2] is to offer a comprehensive review of the application of wireless communication technologies to smart environments by discussing their challenges, future trends, and standardization activities. In [3], various communication technologies, both wired and wireless, are compared, and their suitability for deployment in multiple smart grid applications is evaluated on the basis of specific network requirements. The candidate technologies for a smart grid communication network are also presented and critically discussed in [4]. Moreover, this paper presents a multilayer communication architecture based on IP that integrates heterogeneous technologies by using the ubiquitous sensor network (USN) architecture [5]. The framework includes a decentralized middleware that has to coordinate all the smart system functions, as well as the network management model based on USNs. The integration of heterogeneous technologies is also discussed in detail in [6], where the authors propose a simple and flexible architecture for wireless communications and illustrate how the architecture can be employed. In particular, the paper analyzes the key results obtained from experimental evaluations made using open source software, low cost wireless routers, and open/low cost sensor technology.

Another important challenge is to have secure data transmissions, thus preventing cyber attacks that can block or alter the city functions. A survey on security and privacy issues in smart environments is provided in [7,8], while several solutions are proposed in the literature; e.g., a security framework from wireless sensor networks (WSNs) is proposed in [9]. This framework combines a hierarchical attack detection scheme based on chance discovery and an access control policy. Security is also the focus of [10], which presents a secure IoT architecture. In particular, the authors present an architecture containing four basic IoT blocks that mitigates cyber attacks beginning at the IoT nodes themselves.

The ability to provide heterogeneous services with suitable QoS requirements is investigated in [11–13]. Specifically, [11] presents a survey on the state of the art of cross-layer QoS approaches in WSNs for critical applications. While, in [13], the authors propose four different IP-based IoT network architectures for potential smart city applications, defining the corresponding performance metrics in order to maintain QoS guarantee. As a special case, participatory sensing, as well as its related network architecture and QoS, is separately presented. Differently, in [12], the focus is on access policies of hierarchical WSNs. A new time division multiple access protocol is proposed to improve the quality control of smart cities applications, where diverse traffic is required and loss or delay in data traffic is unacceptable.

Sensors and smart objects composing the smart environment can be subject to power consumption constraints; thus, energy conservation can be essential to extend their lifetime. A comprehensive survey of smart grid-driven approaches in energy-efficient communications and data centres, and of the interaction between smart grid and information and communication infrastructures is provided in [14]. In [15], the problem of energy conservation of smart sensors is considered; the authors propose an approach that is based on a combination of several existing mechanisms of energy saving by addressing more particularly data routing, introducing a hierarchical organization of the network.

To make viable new smart services, it is needed to use the available spectrum more efficiently—both in terms of frequency and with regard to when and where it is needed. A very promising solution is represented by CR technology [16], which allows the support of large-size and time-sensitive multimedia data with limited spectrum resources. In particular, dynamic spectrum access (DSA) enabled by CR technology can provide opportunistic access to unused spectrum, both licensed and unlicensed, making CR technology a necessary component for smart system communication infrastructure. For these reasons, CR is a potentially highly attractive communication technique that, in some contexts, can efficiently address the needs and challenges of a smart city. The rest of the paper focuses on CR as a key enabling communication technology for the smart city.

## 3. Cognitive Radio for Smart City

As stated before, wireless technologies are considered a promising solution for the communication infrastructure of a smart city despite several challenges, such as trade-offs between wireless coverage and capacity as well as limited spectral resources. For this reason, new communication paradigms are needed, and among these, CR networks (CRNs) are highly promising for providing timely wireless communications by utilizing all available spectrum resources.

### 3.1. Cognitive Radio Concepts

The efficiency of spectrum usage in smart systems should be increased, allowing the sharing of radio resources with other systems. CR allows wireless systems acquiring a context-awareness and reconfigure themselves according to the surrounding environments and their own properties [16]. In the same radio resources, two (or more) systems coexist: "primary" and "secondary". Primary system refers to a licensed system with legacy spectrum. This system has the exclusive privilege to access the assigned spectrum. Secondary system refers to the unlicensed cognitive system and can only opportunistically access the spectrum holes which are not used by the primary system.

It is well known that CR methodologies are an interesting research area, but, despite this, until now there has been a lack of wide scale applications of CR techniques. A current attempt is the use of TV white spaces. However, the use of CR schemes in future communication systems for smart cities is mandatory to make spectrum sharing effective among different smart services and communications providers, and foreseen new services [17].

In general, CR technology is based on two main characteristics: "cognitive capability" and "reconfigurability". Cognitive capability refers to the ability to acquire knowledge about the surrounding environment. Reconfigurability enables the secondary network to be dynamically adapted to the radio environment. More specifically, the cognitive radio can be designed to adapt its transmission by means of power control algorithms and/or suitable resource allocation schemes. Two different cognitive approaches can be identified: "Opportunistic" and "Underlay". Following an "opportunistic approach", the secondary system can use portions of radio resources that are unused by the primary system in a given time and space. Hence, the secondary cognitive system uses the radio resources dynamically on an opportunistic and non-interfering basis, exploiting the frequency holes left unused by the primary system. A different cognitive approach is represented by the "underlay cognitive approach". The secondary system is allowed to share the channel simultaneously with the primary, mainly adopting constraints on the power emissions to lower (or avoid) mutual interference. In both cases, the secondary system has to sense the radio channel to estimate which resources are not used among the available ones, or which resources can be used, introducing a limited amount of interference. The cognitive system has to adapt to the changes in the surrounding environment. This means that sensing and reconfigurability must be repeated periodically.

Spectrum sensing is a critical aspect for cognitive systems, especially if the involved devices have low complexity and cannot have multiple radios and powerful processors, like in many smart city environments. Therefore, sophisticated spectrum sensing algorithms cannot be used. Moreover, the sensing duration should be minimized as much as possible to have energy-efficient devices. The trade-off between required resources and sensing accuracy should be addressed for any specific cognitive smart city. Furthermore, to satisfy the QoS requirements of smart city applications, it is needed to have efficient spectrum sharing policies related to medium access control functionalities. More details on these aspects, and a critical discussion related to solutions proposed in the literature is provided in Section 4.

### 3.2. Benefits of Cognitive Radio in Smart City

CR has the potential to flexibly support a wide range of applications and can be useful to deal with several communications challenges in a smart city

- Communication resource availability. CR improves spectrum utilization and communication capacity to support large-scale data transmissions. Indeed, the unlicensed spectrum (*i.e.*, Industrial, Scientific, and Medical, ISM) mainly used in local area connections is becoming dramatically crowded and interfered, while other licensed frequency bands are fixedly assigned and utilized in an inefficient way. In addition, the application of CR can also alleviate the burden of purchasing licensed spectrum for utility providers. CR uses the existing spectrum through opportunistic access to the licensed bands without interfering with the licensed users. CR determines the spectrum portions unoccupied by the licensed users—known as spectrum holes or white spaces—and allocates the best available channels for communicating.
- Heterogeneity. Heterogeneous communication technologies have to be integrated to provide reliable and efficient access to the system elements in different environments. As a consequence, devices should be able to acquire context awareness and to reconfigure themselves. Hardware reconfigurability can help to manage communications in areas where different technologies are present.
- Quality of Service. Communications over white spaces can provide dedicated low-latency communications for critical data.
- Energy consumption. CR can be used to reduce power consumption, and hence to have energy efficient systems, by sensing the environment and then adaptively adjusting the transmission power, avoiding energy waste.

### 3.3. Related Literature Review

Supporting smart cities through CR communications is becoming an interesting area of research, especially in the context of smart grids and smart vehicular networks.

Several surveys on smart systems based on CR can be found in the literature. The application of CR to future generation networks is provided in [18]. The authors in [19] focus on the main features of cognitive vehicular networks, providing an overview of the state of the art, especially in terms of spectrum sensing and open research problems. Similarly, in [17,20], application scenarios, motivations, and challenges of using CR for smart grids are reviewed. Moreover, the authors in [17,20] provide a survey of possible architectures and spectrum sensing techniques. CR technology is also reviewed in [21,22] as a possible solution for implementing effective smart grid networks.

General communication architectures for smart grids based on CR are discussed in several papers, as in [23–25]. In particular, [23] provides an overview of the current state of communication technologies for smart grids, and then the possibility of applying CR is discussed together with a high level network architecture based on IEEE 802.22 standard. In [24,25], the authors present a CR-based communication architecture, organized in three tiers depending on the service area (*i.e.*, home, neighbourhood, wide). Moreover, in [24], dynamic spectrum access and sharing in each subarea are considered, evidencing the necessity of joint resource management in different subareas in order to achieve network scale performance optimization.

Despite the benefits of CR, when many unplanned networks simultaneously access a common pool of frequency channels, high background interference occurs. In [26], a beamforming technique based on minimum mean squared error (MMSE) is proposed to mitigate the interference in CR systems based on the IEEE802.22 standard for smart meter. Another problem that arises in CR is security of sensed data, this is discussed in [27], where a two-stage scheme for defence against spectrum sensing data falsification attacks is proposed.

CR is also investigated as a viable solution to have efficient WSNs, which are considered to be one of the best solutions as a monitoring platform for many smart systems. Dynamic and opportunistic spectrum access capabilities of CR can address many of the unique requirements and challenges of WSNs: propagation conditions, heterogeneous spectrum characteristics varying over time and space, reliability and latency requirements, and energy constraints. For example, in [28], spectrum-aware WSNs are proposed to overcome spatio-temporally varying spectrum characteristics and harsh environmental conditions for smart grid applications: potential applications, challenges, and protocol design principles are reviewed. The performance of a WSN in a smart meter communication system in terms of average service time and average waiting time is evaluated in [29], where an overlay cognitive implementation strategy is adopted, treating the WiFi system as the primary user.

Scheduling problems in CR smart grids are addressed in [30–32], where different priority-based solutions are proposed considering the heterogeneity of the traffic generated by the specific environments.

CR technology is also investigated as a viable solution for mitigating the spectrum shortage that can arise from the use of some key technologies that could enable efficient smart city environments. In particular, HetNets and M2M communications are two paradigms particularly suited for future smart cities, but their effectiveness is related to the capability of exploiting the available spectrum in an efficient way. In the remaining part of this paper, we focus on CR technology applied to these new and promising communication paradigms.

### 3.4. Cognitive M2M

Smart systems are characterized by a large diversity of devices and machines that have to be interconnected and able to exchange information autonomously in order to make the environment smart. This makes M2M communications a dominant paradigm, especially in contexts such as in-home applications, vehicular telematics, healthcare, and public safety [33–36]. However, M2M communications present multiple differences with respect to current Human-to-Human (H2H) information production, processing, and exchange. M2M communications are characterized by massive transmissions, small bursty traffic, low power, low cost, and low mobility. Moreover, while H2H communications access the network following resource scheduling policies based on traditional bandwidth request mechanisms, M2M communications need to be separately considered. Indeed, these are characterized by a high number of nodes that exchange low data rate information. Hence, the spectrum usage is very limited and short in time. However, the main challenge for M2M communications is due to the high number of involved devices, which excessively increases the access requests to the network and, without countermeasures, gives rise to network congestion. Indeed, even if each device transmits only a very small amount of data, the huge number of connected devices will dramatically increase the signaling overhead, thus leading to a network overload and deteriorating the QoS of other H2H applications. A viable solution is to develop an autonomous M2M communication system that shares the resources with other H2H communication systems without causing congestion. This is a new paradigm, called Cognitive M2M (CM2M) communications [37,38]. The exploitation of CR technology for setting up M2M communications is a promising trend. Indeed, a cognitive approach can lead to the definition of a scenario where M2M nodes are constrained to send data in specific bands and time intervals, reducing the congestion issues. CM2M communications concern with the presence of a primary H2H communication system that coexists with a secondary CM2M system in an almost transparent way, avoiding interference and introducing low performance degradation thanks to the cognitive approach. The cognitive engine of an M2M network can operate in a distributed manner (among all the devices of the network) or in a centralized manner at the network gateway, which is in general supposed to be a more powerful node. A more detailed discussion on CM2M solutions is provided in Section 4.

### 3.5. Cognitive HetNets

As stated in Section 2, one of the main characteristics of the communication architecture of a smart system is that, in general, it spreads over large geographical areas and, hence, requires an infrastructure characterized by a multilayer network with different access points in order to provide network access in local areas for in-home applications up to wide areas, for communications with command and control centres. A promising technology is represented by the HetNets deployment, which is a key networking paradigm in 5G mobile networks based on the idea of increasing the number of access points with a high diffusion of short-range, low-power, and low-cost base stations (BSs) overlapped to the main 5G macro-cellular network infrastructure [39,40]. This allows to provide high data rates, to offload traffic from the macro cell, and to provide dedicated capacity to smart cities [41,42]. In HetNets, a broad variety of cells, such as micro, pico, metro, and femto cells, as well as advanced wireless relays and distributed antennas, can be deployed practically anywhere. Since the ranges are short due to low transmission power, the small cells are typically deployed in close proximity to the users and can be used for creating hot-spots for access to smart services and applications, increasing in-home connectivity, providing accurate localization, and integrating macro-cellular coverage. However, the deployment of two overlapped layers of cells (*i.e.*, macro and small cells) requires networks able to self-organize and to manage the inter-cell interference. Indeed, interference mitigation between the two network layers is one of the key issues to be solved before the effective application of the HetNet concept. The introduction of small cells in current cellular networks is based mainly on coordinated approaches where the small cell is directly installed by the network operator, and it is possible to adopt coordinated resource allocation strategies, thus avoiding inter-cell interference. However, in smart city environments, small cells will be deployed *ad-hoc* in a flexible and scalable way, depending on the system's needs and, hence, without a coordination with the network operators. In this way, a more cost-effective communication system will be achieved, reducing the costs for RF planning, site acquisition, and efficient backhauling [39]. This trend is in general expected for future 5G mobile networks where the number of small cell nodes will increase significantly, and many user-deployed small cells will be used in homes, small offices, and enterprises. If coordination among macro and small cells in the resource assignment is unavailable, the concept of Cognitive HetNets (CHetNets) [43,44] assumes high relevance. The macrocell is the primary system that has higher priority on the resource usage and the cognitive small cell (CSC) represents the secondary system that has lower priority and should transmit without affecting the primary system reception. The CSC radio access network must be equipped with a cognitive engine able to sense the environment and to adapt transmissions by means of DSA schemes and advanced signal processing methods. Some examples of cognitive approaches used by the small cell cognitive engine (SMCE) are provided in Section 4.

### 3.6. Cognitive Communication Architecture for Smart City

As a conclusion of this section, we want to suggest how CHetNets and CM2M communications could be integrated in the communication infrastructure of a smart environment, proposing a high level architecture based on a 4G/5G cellular network. Indeed, even if different technologies will be used to support smart city communications, it is reasonable to assume that cellular networks will have a predominant role for multiple reasons—such as the performance they can guarantee in terms of data rate, delay and security, coverage, and the fact that they already exist and are widespread so can be used anywhere and at any time without extra costs to deploy a smart environment. The drawback is that the users of the cellular networks are continuously increasing in number and the demand of capacity is expected to grow exponentially in the near future. This could determine congestion and a consequent loss of performance of the smart system. To resolve these challenges, it is essential to adopt a network infrastructure that can efficiently integrate multiple disruptive wireless technologies and enable interworking of existing and deployed technologies. For this reason, we consider CR, HetNets, and M2M technologies.

The architecture we propose is represented in Figure 1. This is able to provide both local and backbone connectivity to the smart city in an efficient way. We exploit the multi-layer nature of the communication infrastructure of a smart city, proposing a model where the cellular network, with its main macro-cellular coverage provides wide area connections, while in areas where smart services and applications are concentrated, the communications are supported by small cells deployed *ad-hoc*. Hence, where peaks of traffic are expected, the radio access is guaranteed by dedicated cognitive small cells (DCSC) that provide service only to the smart city users. Moreover, devices in close proximity can directly communicate (device-to-device, D2D) among them, thus offloading traffic from the BS. This enables the creation of another layer in the smart system communication architecture which is represented by the CM2M layer, thus supporting the massive diffusion of connected devices usually characterizing a smart city. In the CM2M layer, machines can communicate among themselves by opportunistically using the spectrum both in underlay or overlay mode. Then, the data collected and processed by the CM2M network are forwarded toward the control centre using gateway nodes that have two air interfaces and are able to connect with the DCSC. This follows the capillary network concept [45].

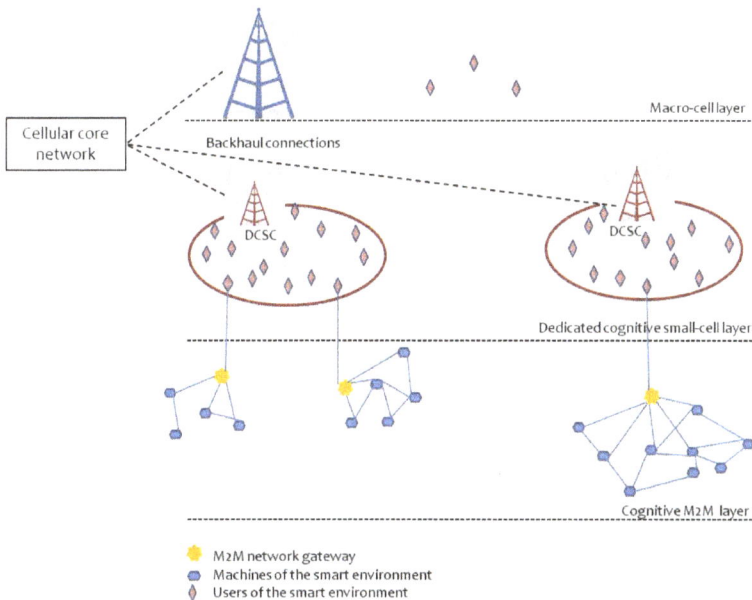

**Figure 1.** Cognitive communication architecture for smart city. M2M: Machine-to-Machine; DCSC: dedicated cognitive small cell.

To efficiently work CHetNets and CM2M, networks have to resort to suitable cognitive solutions, for which a critical review of possible approaches is provided in the next section.

## 4. Cognitive Solutions for HetNets and M2M Communications

This section deals with some possible solutions to be applied to CHetNets and CM2M communications in a smart city, where one of the main challenges related to the use of CR technologies is spectrum sensing. Indeed, as stated previously, when sensing involves low complexity devices as in a smart environment, it is necessary to find a suitable trade-off between the amount of resources requested by the sensing operation (in terms of computational complexity and energy) and the sensing accuracy (in terms of the probability of correct detection and sensing time). As a consequence, depending on the cognitive node capabilities, different CR approaches should be selected. For this reason, we critically review different solutions distinguishing between two classes of cognitive systems that differ in the speed of changing of the primary system resource usage, and hence for the complexity of sensing operation:

1. long-term (*i.e.*, seconds) cognitive systems;
2. short-term (*i.e.*, milliseconds) cognitive systems.

In the first case, the primary and the secondary systems share portions of the available spectrum (sub-bands) indicated as $B_i$ in Figure 2, whose occupancy changes due to significant modifications in the environment. This is the case of future 5G cellular systems, where the need for wide frequency bands and the unavailability of large free spectrum portions will lead to the dynamic aggregation of multiple, even non-contiguous, sub-bands ("carrier aggregation") [46] to satisfy the requested capacity. Hence, each primary cellular system adaptively changes the number of used sub-bands depending on the traffic load and the interference level with neighbour cells. Together with carrier aggregation, future 5G systems also foresee the exploitation of unlicensed spectrum portions (*i.e.*, the ISM spectrum) to increase the capacity of small cells in HetNets, thus reducing the interference generated towards the macrocell. The use of unlicensed spectrum is subject to its availability and therefore sensing must be repeated periodically to find unused sub-channels [47]. Another interesting case is represented by the IEEE 802.22 standard for wireless regional area networks, which is based on the opportunistic usage of available portions of TV spectrum (TV white spaces) that are unused in a given area and in a given time. In all these cases, cognitive secondary systems must be able to detect the use of a given sub-band performing suitable spectrum sensing. The sub-bands where no activity of the nearby primary system is detected are used to communicate with the secondary users. In this case, it is reasonable to assume that the primary system sub-bands occupancy does not change frame-by-frame, but it is performed periodically when the network load changes significantly, or when new cells are activated in the same area.

Short-term cognitive approaches are more challenging in terms of sensing because they foresee the exploitation of the smallest resource units (RUs) in which a sub-band can be divided (indicated as $RU_{i,j}$ in Figure 2), and these are allocated frame-by-frame requiring instantaneous sensing and decisions. Indeed, current and next generation wireless communication systems are characterized by a high flexible and dynamic resource usage. Therefore, the available sub-bands are divided into small RUs (For example, the Physical Resource Blocks in LTE-A.) that are instantaneously allocated to users in a dynamic way depending on the users' requests and channel propagation conditions.

In summary, the main difference between the two approaches is the time scale. In the first case, we can assume that the transmission opportunities last for some frames, while in the second case the resource assignment changes frame-by-frame. The short-term cognitive approach is more efficient because it works with a higher granularity of the resources and, hence, it permits to exploit all the resources left unused by the primary system. On the other hand, this approach is very challenging for the sensing phase, which must be very quick and repeated with high frequency. Conversely, in long-term operations, sensing requirements can be relaxed: sensing duration is longer and frequency of sensing is lower.

Finally, in the short-term cognitive approach, the secondary network must use the same access technology of the primary network and has to be synchronized with it. Conversely, with the long-term cognitive approach, different technologies can be used in the secondary network, hence, it can also operate on sub-bands allocated to different primary systems, including licensed and unlicensed spectrum.

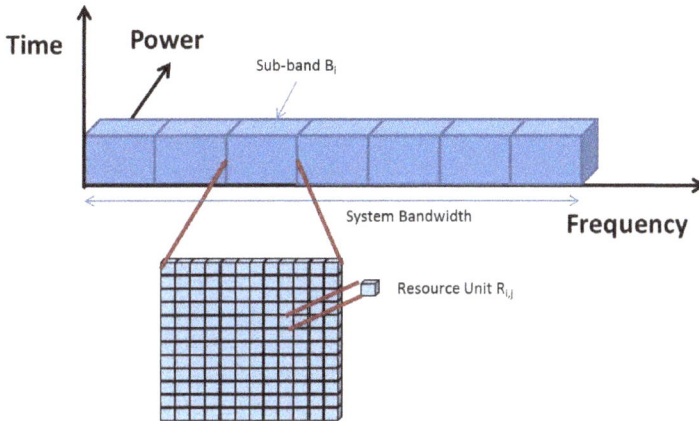

**Figure 2.** Spectrum Opportunities.

As a first consideration, we can state that in the smart city architecture described in Section 3.6, the long-term cognitive approaches are most suitable for the low complexity devices that compose the CM2M networks, while cognitive small cells have the capabilities of supporting both types of approaches. A critical discussion is provided in the next section.

*4.1. Long-Term Cognitive Approaches*

In long-term cognitive approaches, the main problem is making a cognitive device aware—in an autonomic manner—of which frequency sub-bands are not used by the primary system, and hence, able exploit the transmission opportunities. Therefore, the secondary system plans periodic sensing intervals during which it measures the energy in the whole sub-bands in order to detect primary system activity.

Several spectrum sensing approaches can be used among those proposed in the literature [48]. The spectrum sensing algorithms can be classified into blind detection and feature detection. The former is applicable to all the primary signals, even unknown, since prior information on the signal is not required. The most-known and simplest blind method is Energy Detector, but it has the drawback that it suffers heavy degradation in the case of noise uncertainty and in a low signal-to-noise ratio (SNR) regime. Differently, feature detectors such as matched filter, waveform-based, and cyclostationary sense specific characteristics of the primary signals that must be known "a priori". These methods can provide accurate sensing performance even in the presence of noise. In particular, great attention has been devoted to the methods based on the cyclostationarity property of the signal induced by the Cyclic Prefix (CP) in Orthogonal Frequency Division Multiplexing (OFDM) signals, such the ones used in most of the current wireless systems (*i.e.*, LTE-A, WiFi, WiMAX). Different sub-optimal methods exploiting cyclostationarity have been proposed and evaluated to limit the computational complexity. In smart city environments, the selection of the sensing scheme should be suitably driven by the desired trade-off between implementation complexity, estimation time, and performance in terms of false alarm and detection probability. In particular, when low complexity devices are involved (*i.e.*, sensors), low complexity algorithms should be preferable—adopting different strategies, such as distributed sensing [49], to improve the reliability of the produced output.

In the literature, some examples of long-term cognitive approaches can be found. In [50], the authors proposed a selective method that permits the secondary user to exploit not only the sub-bands left unused by the primary system (Opportunistic approach) but also the sub-bands that result underutilized by adopting suitable subcarrier. This solution can be useful either in HetNets or M2M communications. However, underlay communications are particularly suitable for M2M applications characterized by short range communications, and therefore low power emissions. In particular, depending on the primary systems' signal strength, the CM2M network is able to determine the maximum power that can be used without causing high interference. This issue was considered in [51], where the authors analyzed the feasibility of implementing a CM2M network by using primary cellular bands. A hierarchical network structure was proposed, where cluster heads gather M2M traffic and forward it to the primary cellular network. In addition, the cluster head—on the basis of a previous sensing phase—estimates the maximum allowable power for M2M communications and broadcasts this information to all M2M devices within its area. In [38,52], CM2M communications have been investigated to be applied in smart environments. In [38], the authors designed a cognitive medium access control protocol, based on packet reservation multiple access. The protocol is centralized and utilizes a specialized frame structure for supporting the coexistence of the CM2M network with the primary H2H network. In this case, spectrum sensing is not performed by each machine, but a low-cost dynamic spectrum access solution realized in the form of master–slave operation is considered. It means that only the gateway of the M2M network, which is supposed to be a more powerful device, performs sensing and then makes a decision on the resource access. The gateway sends an enabling signal after obtaining a vacant channel and then other devices are allowed to transmit in vacant channels. The opposite approach was considered in the M2M network presented in [52], where smart objects were able to connect to each other in a distributed manner so that the sensing information could hop from one node to other nodes until it reached the gateway successfully. The focus of the paper was on improving the sensing capabilities of the system by using a differential evolution algorithm that exploits the sensing results coming from multiple objects. A HetNet deployment was proposed in [53] for a smart grid. It is based on the integration of heterogeneous access points and DSA. In this case, the selection of the best access point and of the spectrum resources is centralized. Indeed, a network controller uses average statistics to assign the BS to each device, and a spectrum manager maintains a database of available and leased open spectrum without the need for spectrum sensing operations. Differently [54] mainly focused on the capability of a secondary access point to detect the activity of the primary macrocell that operates following a carrier aggregation policy. Hence, the small cell performs suitable spectrum sensing based on low complexity cyclostationary detectors to understand to understand in an autonomic manner which frequency bands are available.

### 4.2. Short-Term Cognitive Approaches

Short-term cognitive approaches represent a very promising solution due to their specific features. However, many challenging issues have to be overcome to make viable their use in future smart systems. A short-term cognitive algorithm has to operate quickly in order to be aware of the spectrum resource usage by the primary systems in real time and then has to make quick and efficient access decisions for the secondary devices. Moreover, a short-term cognitive approach works in different modes depending which link—*i.e.*, UpLink (UL) or DownLink (DL)—is considered. The access point of a cognitive secondary system listens the environment in order to acquire knowledge about the UL transmission of the primary system. This information can be used to allocate the resources in the successive UL transmission. The DL knowledge, however, must be acquired by the terminals during a given time interval and sent back to the access point, which performs the allocation in the following DL transmission. This introduces a certain latency between the acquisition of the information and its use, which must be accurately evaluated in accordance with the working speed of the scheduler. This could be critical in Time Division Duplexing (TDD) systems where the cognitive terminal has to wait the UL sub-frame before sending the information back to the access point. Despite this, TDD based systems are attractive for the use of short-term cognitive methods because for them the channel reciprocity is applicable, that allows to keep valid the channel estimates performed in the DL also for the UL and viceversa. The main differences between long-term and short-term cognitive approaches are reported in Table 1. It is evidenced that short-term approaches are mainly suitable for CHetnNets to be used in the access layer of a smart city, while is not suitable for the sensing layer; however, there is a special case based on a partial knowledge of the primary system scheduling information that can also be suitable for CM2M applications, as detailed later.

**Table 1.** Cognitive Approaches. CM2M: Cognitive machine-to-machine; CHetNet: Cognitive heterogeneous networks.

| Cognitive approaches features | Long-Term | Short-Term |
|---|---|---|
| Smart city communication layer | local and access layer (CM2M and CHetNets) | access layer (CHetNets) |
| Sensing period | several frames | scheduling period |
| Transmission opportunity | sub-bands | Resource Units |
| Technical challenges | suitable trade off: cost-accuracy of sensing | fast sensing feedback information joint resource allocation |
| Spectrum efficiency | Low | High |
| Secondary Network Requirements | no | synchronization and legacy terminal |
| Distributed sensing | yes | no |

Most of the research activity concerning short-term cognitive approaches focuses on finding efficient (e.g., optimal) resource allocation strategies to minimize the interference of cognitive small cells towards the primary macrocell system without losing performance for the secondary devices. The resource allocation schemes benefit from the flexibility given by multiuser diversity; when working on a RU basis, different secondary users have different spectrum opportunities available. Indeed, depending on their positions, the secondary users receive and produce different levels of interference from/towards the primary users. This means that if a primary user is communicating by using a given resource, a proximity secondary user senses this resource as occupied, but another secondary user that is far from the primary one senses the resource as free and can use it for communicating. To use this multiuser diversity property, the spectrum sensing must perform joint spatial–temporal resource detection.

Awareness of the macrocell resource usage can be achieved in two different ways: by receiving the scheduling information from the macrocell BS, or by actively sensing the environment. In the first case, a limited level of signaling exchange among the BSs is required, while in the second case, the information is acquired using sensing procedures without any type of information exchange. The first method is not completely cognitive because there is only limited cooperation between the BSs, but the information exchange represents only one-sided cooperation, and interference management mainly relies on cognitive approaches. In addition, use of the scheduling information sent by the macrocell has the drawback that the multiuser diversity remains unused. Indeed, when a resource is declared as used by the primary system, it is considered busy by all secondary users and cannot even be used by a user located far away from the primary system, which could in fact operate without interference. The second method is completely cognitive, but presents more challenges in the sensing phase, not only in terms of accuracy of the results, but also in terms of latency between acquisition of the context awareness and its use and channel reciprocity between the UL and the DL. In some cases, hybrid techniques can represent a viable solution whenever limited knowledge of the network is available, and therefore only partial coordination among the cells is possible.

The method proposed in [55] first detects channel occupation by estimation of the energy in the UL sub-channels, and then allocates the sub-channels with the lowest interference signatures to the small cell users. The hypothesis here is that the same resource scheduling process is used in both UL and DL transmissions. However, it is more likely that the DL and the UL are characterized by asymmetric traffic, and thus adopt different resource allocation policies. Therefore, UL sensing cannot be used for DL allocation. An alternative approach is proposed in [43] by considering a hybrid sensing scheme in which the scheduled macrocell BS information is available at the small cell BS (SBS) in order to increase the spectrum sensing accuracy. In this way, the secondary system finds more spectrum opportunities by identifying nearby macrocell users. Inter-layer interference can also be limited or prevented using optimal power allocation and using underlay cognitive approaches, as in [56], where the SBS senses both the UL and DL of the macrocell to be aware of both the resource occupancy and nearby macrocell users. The algorithm then adapts the power on each resource element to maximize the achievable small cell throughput and fulfill the macrocell users' outage constraints.

Exploitation of multi-antenna technologies is an additional opportunity to use information related to the primary and secondary users' position in order to allow co-channel frequency allocation between primary and secondary systems. A useful approach is to use beamforming at the small cell transmitter in order to maximize the secondary system performance while the interference on the primary system receiver is minimized [57]. This operation is named Cognitive Beamforming (CB) and requires complex numerical solutions and the knowledge of all propagation channels. This could be impracticable in actual scenarios, as the two systems operate in an independent mode that prevents each of them from detecting the presence of the other. Hence, it is necessary to resort to sub-optimal solutions that can work with a partial knowledge of the channel state information and with some information exchange between primary and secondary networks. A viable low complexity CB approach is based on the exploitation of the direction of arrival (DoA): the secondary system can transmit avoiding interference on the primary user by placing nulls in its direction. However, the knowledge of DoA of multiple signals can be challenging in multipath propagation channels and strongly depends on the number of antenna elements used at the receiving end. In [58], a method is proposed for CHetNets based on DoA estimation and zero forcing beamforming that focuses on problems that arise in actual propagation channels. This method can be applied either in UL or DL, assuming a correlated spatial information among the two links. Finally, CB methods can be combined with opportunistic resource allocation algorithms that assign each *RU* to the secondary user that is sufficiently far from the corresponding primary. Thus, the interference can be minimized with beamforming. An example based on DoA information is sketched in Figure 3. In this method, the knowledge of the DoA can be acquired by the SBS during the UL transmission and then used either for the UL reception or for the DL transmission. However, for the DL transmission, two issues must be taken into account: the channel reciprocity in the spatial information between DL and UL and the need of the SBS to know the scheduling map (*i.e.*, the resource assignment to the users) of the primary system. This could be achieved with a limited amount of information exchange among the BSs.

A context where the knowledge of the scheduling maps of the primary system is particularly interesting is the CM2M. In addition to the problems related to the sensing delay, in CM2M networks, spectrum sensing operation requires digital processing and energy consumption that can be unaffordable. Indeed, the machines involved in autonomous M2M communications in a smart city are usually characterized by low complexity and low cost nodes, with reduced computation capabilities and with stringent requirements on battery life (for example, sensor nodes or metering devices). Moreover, the large number of nodes could occur in a huge amount of signaling required to exchange sensing reports. As already stated, in CM2M communications, spectrum sensing can be a very hard task. A viable solution is proposed in [59], where a CM2M network aims to exploit the time-frequency holes left in the H2H communication frame for implementing an independent network. Toward this goal, the proposed system uses the in-band signaling broadcast by the primary system to discover the unused spectrum parts for the cognitive capability of the M2M network. Indeed, a specific field within the H2H is used for notifying in broadcast to the users on the spectrum allocation. In particular, [59] proposes a novel M2M MAC technique suitable to support M2M communications, with the aim of allowing the multiple access to secondary users while avoiding interferences to the already-planned primary network; the secondary CM2M devices are supposed to have a legacy interface so that the in-band signaling of the overlaid network can be exploited in a suitable way. In addition, it is important to underline that using the scheduling MAPs to detect idle RUs has the disadvantage that the multiuser diversity is not exploited, and when a resource is used by the primary system it cannot be used by the secondary. Neither can it be used by a secondary user that is far from the primary system, and hence it could operate without interference.

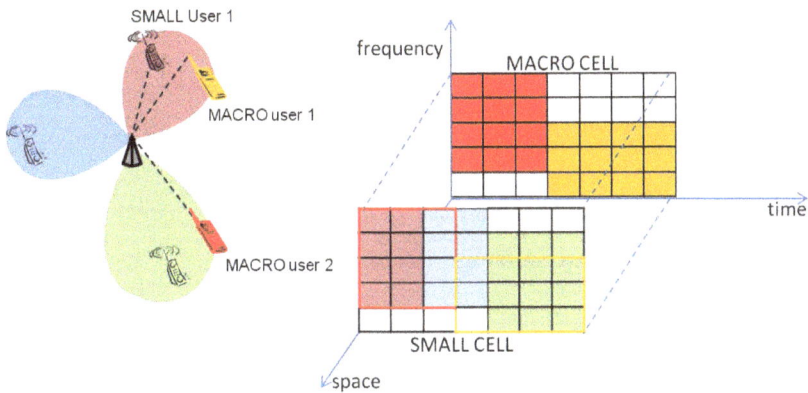

**Figure 3.** Cognitve Beamforming and Resource allocation.

Table 2 presents a summary of characteristics of the short-term cognitive approaches based on sensing or on the knowledge of the primary system's scheduling map.

**Table 2.** Short-term cognitive approaches comparison.

| Cognitive approaches features | Sensing | Signaling (Scheduling maps) |
|---|---|---|
| Computational Complexity | high | low |
| Challenges | real-time sensing | scheduling map availability |
| Applications | HetNets | HetNets and M2M |
| Spectrum Efficiency | high | medium |

## 5. Conclusions

In the near future, the quality of life of billions of people will depend on the capability to efficiently use information and communication technologies to build smart environments able to save energy, reduce harmful emissions, improve health care and increase citizen security. A smart city is based on a functional integration of software systems, network infrastructures, heterogeneous user devices, and collaboration technologies. Among these components, a high performance, reliable, robust, secure, and scalable communication architecture plays a crucial role for the successful operation of a smart environment. The capability of wireless communication systems to satisfy the smart city requirements is related to their capacity to efficiently exploit the spectrum resources, and to introduce new communication paradigms particularly suitable for this kind of environment. For this reason, in this paper, after a review of communication challenges and solutions, cognitive ratio technology has been introduced as a key feature to make effective smart environments, and in particular to mitigating the potential spectrum shortage arising from the use of HetNets and M2M communications. Related literature has been also reviewed and critically discussed in order to highlight the effectiveness of these key technologies in a smart city environment.

**Conflicts of Interest:** The authors declare no conflict of interest.

## References

1.  Ma, R.; Chen, H.H.; Huang, Y.R.; Meng, W. Smart Grid Communication: Its Challenges and Opportunities. *IEEE Trans. Smart Grid* **2013**, *4*, 36–46.
2.  Fan, Z.; Kulkarni, P.; Gormus, S.; Efthymiou, C.; Kalogridis, G.; Sooriyabandara, M.; Zhu, Z.; Lambotharan, S.; Chin, W.H. Smart Grid Communications: Overview of Research Challenges, Solutions, and Standardization Activities. *IEEE Commun. Surv. Tutor.* **2013**, *15*, 21–38.
3.  Kuzlu, M.; Pipattanasomporn, M. Assessment of communication technologies and network requirements for different smart grid applications. In Proceedings of the IEEE PES Innovative Smart Grid Technologies (ISGT), Washington, DC, USA, 24–27 February 2013; pp. 1–6.
4.  Zaballos, A.; Vallejo, A.; Selga, J.M. Heterogeneous communication architecture for the smart grid. *IEEE Netw.* **2011**, *25*, 30–37.
5.  ITU. *Ubiquitous Sensor Networks (USN)*; ITU-T Technology Watch Briefing Report Series, No.4; ITU-T: Geneva, Switzerland, February 2008.
6.  Avelara, E.; Marquesa, L.; dos Passosa, D.; Macedob, R.; Diasa, K.; Nogueirab, M. Interoperability issues on heterogeneous wireless communication for smart cities. *Comput. Commun.* **2015**, *58*, 4–15.
7.  Komninos, N.; Philippou, E.; Pitsillides, A. Survey in Smart Grid and Smart Home Security: Issues, Challenges and Countermeasures. *IEEE Commun. Surv. Tutor.* **2014**, *16*, 1933–1954.
8.  Finster, S.; Baumgart, I. Privacy-Aware Smart Metering: A Survey. *IEEE Commun. Surv. Tutor.* **2015**, *17*, 1088–1101.
9.  Wu, J.; Ota, K.; Dong, M.; Li, C. A Hierarchical Security Framework for Defending Against Sophisticated Attacks on Wireless Sensor Networks in Smart Cities. *IEEE Access* **2016**, *4*, 416–424.
10. Chakrabarty, S.; Engels, D.W. A secure IoT architecture for Smart Cities. In Proceedings of the 13th IEEE Annual Consumer Communications Networking Conference (CCNC), Las Vegas, NV, USA, 9–12 January 2016; pp. 812–813.
11. Al-Anbagi, I.; Erol-Kantarci, M.; Mouftah, H.T. A Survey on Cross-Layer Quality-of-Service Approaches in WSNs for Delay and Reliability-Aware Applications. *IEEE Commun. Surv. Tutor.* **2016**, *18*, 525–552.
12. Alvi, A.N.; Bouk, S.H.; Ahmed, S.H.; Yaqub, M.A.; Sarkar, M.; Song, H. BEST-MAC: Bitmap-Assisted Efficient and Scalable TDMA-Based WSN MAC Protocol for Smart Cities. *IEEE Access* **2016**, *4*, 312–322.
13. Jin, J.; Gubbi, J.; Luo, T.; Palaniswami, M. Network architecture and QoS issues in the internet of things for a smart city. In Proceedings of the International Symposium on Communications and Information Technologies (ISCIT), Gold Coast, Australia, 2–5 October 2012; pp. 956–961.
14. Erol-Kantarci, M.; Mouftah, H.T. Energy-Efficient Information and Communication Infrastructures in the Smart Grid: A Survey on Interactions and Open Issues. *IEEE Commun. Surv. Tutor.* **2015**, *17*, 179–197.
15. Imen, B.; Mahmoud, P.A. Hierarchical organization with a cross layers using smart sensors for intelligent cities. In Proceedings of the SAI Intelligent Systems Conference (IntelliSys), London, UK, 10–11 November 2015; pp. 446–451.
16. Haykin, S. Cognitive radio: Brain-empowered wireless communications. *IEEE J. Sel. Areas Commun.* **2005**, *23*, 201–220.
17. Khan, A.A.; Rehmani, M.H.; Reisslein, M. Cognitive Radio for Smart Grids: Survey of Architectures, Spectrum Sensing Mechanisms, and Networking Protocols. *IEEE Commun. Surv. Tutor.* **2016**, *18*, 860–898.
18. Niyato, D.; Hossain, E. Cognitive radio for next-generation wireless networks: An approach to opportunistic channel selection in ieee 802.11-based wireless mesh. *IEEE Wirel. Commun.* **2009**, *16*, doi:10.1109/MWC.2009.4804368.
19. Felice, M.D.; Doost-Mohammady, R.; Chowdhury, K.R.; Bononi, L. Smart Radios for Smart Vehicles: Cognitive Vehicular Networks. *IEEE Veh. Technol. Mag.* **2012**, *7*, 26–33.
20. Gungor, V.C.; Sahin, D. Cognitive Radio Networks for Smart Grid Applications: A Promising Technology to Overcome Spectrum Inefficiency. *IEEE Veh. Technol. Mag.* **2012**, *7*, 41–46.

21. Kouhdaragh, V.; Tarchi, D.; Coralli, A.V.; Corazza, G.E. Cognitive Radio based Smart Grid Networks. In Proceedings of the 24th Tyrrhenian International Workshop on Digital Communications—Green ICT (TIWDC), Genoa, Italy, 23–25 September 2013; pp. 1–6.
22. Sum, C.S.; Harada, H.; Kojima, F.; Lan, Z.; Funada, R. Smart utility networks in TV white space. *IEEE Commun. Mag.* **2011**, *49*, 132–139.
23. Gao, J.; Wang, J.; Wang, B.; Song, X. Cognitive radio based communication network architecture for smart grid. In Proceedings of the International Conference on Information Science and Technology (ICIST), Changsha, Hubei, China, 23–25 March 2012; pp. 886–888.
24. Yu, R.; Zhang, Y.; Gjessing, S.; Yuen, C.; Xie, S.; Guizani, M. Cognitive radio based hierarchical communications infrastructure for smart grid. *IEEE Netw.* **2011**, *25*, 6–14.
25. Vineeta, N.; Thathagar, J.K. Cognitive radio communication architecture in smart grid reconfigurability. In Proceedings of the 1st International Conference on Emerging Technology Trends in Electronics, Communication and Networking (ET2ECN), Surat, Gujarat, India, 19–21 December 2012; pp. 1–6.
26. Chang, S.; Nagothu, K.; Kelley, B.; Jamshidi, M.M. A Beamforming Approach to Smart Grid Systems Based on Cloud Cognitive Radio. *IEEE Syst. J.* **2014**, *8*, 461–470.
27. Basharat, M.; Ejaz, W.; Ahmed, S.H. Securing cognitive radio enabled smart grid systems against cyber attacks. In Proceedings of the First International Conference on Anti-Cybercrime (ICACC), Riyadh, Kingdom of Saudi Arabia, 10–12 November 2015; pp. 1–6.
28. Bicen, A.O.; Akan, O.B.; Gungor, V.C. Spectrum-aware and cognitive sensor networks for smart grid applications. *IEEE Commun. Mag.* **2012**, *50*, 158–165.
29. Yang, H.C.; Zhang, D.; Kong, X.; Jia, H. Performance Analysis of Cognitive Transmission in Dual-Cell Environment and its Application to Smart Meter Communications. In Proceedings of the Seventh International Conference on Broadband, Wireless Computing, Communication and Applications (BWCCA), Victoria, BC, Canada, 12–14 November 2012; pp. 40–45.
30. Huang, J.; Wang, H.; Qian, Y.; Wang, C. Priority-Based Traffic Scheduling and Utility Optimization for Cognitive Radio Communication Infrastructure-Based Smart Grid. *IEEE Trans. Smart Grid* **2013**, *4*, 78–86.
31. Siya, X.; Lei, W.; Zhu, L.; Shaoyong, G.; Xuesong, Q.; Luoming, M. A QoS-aware packet scheduling mechanism in cognitive radio networks for smart grid applications. *China Commun.* **2016**, *13*, 68–78.
32. Yu, R.; Zhong, W.; Xie, S.; Zhang, Y.; Zhang, Y. QoS Differential Scheduling in Cognitive-Radio-Based Smart Grid Networks: An Adaptive Dynamic Programming Approach. *IEEE Trans. Neural Netw. Learn. Syst.* **2016**, *27*, 435–443.
33. Wan, J.; Li, D.; Zou, C.; Zhou, K. M2M Communications for Smart City: An Event-Based Architecture. In Proceedings of the IEEE 12th International Conference on Computer and Information Technology (CIT), Chengdu, China, 27–29 October 2012; pp. 895–900.
34. Elmangoush, A.; Coskun, H.; Wahle, S.; Magedanz, T. Design aspects for a reference M2M communication platform for Smart Cities. In Proceedings of the 9th International Conference on Innovations in Information Technology (IIT), Abu Dhabi, United Arab Emirates, 17–19 March 2013; pp. 204–209.
35. Skouby, K.E.; Lynggaard, P. Smart home and smart city solutions enabled by 5G, IoT, AAI and CoT services. In Proceedings of the International Conference on Contemporary Computing and Informatics (IC3I), Mysore, India, 27–29 November 2014; pp. 874–878.
36. Datta, S.K.; Bonnet, C. Internet of Things and M2M Communications as Enablers of Smart City Initiatives. In Proceedings of the 9th International Conference on Next Generation Mobile Applications, Services and Technologies, Cambridge, UK, 9–11 September 2015; pp. 393–398.
37. Zhang, Y.; Yu, R.; Nekovee, M.; Liu, Y.; Xie, S.; Gjessing, S. Cognitive machine-to-machine communications: Visions and potentials for the smart grid. *IEEE Netw.* **2012**, *26*, 6–13.
38. Aijaz, A.; Aghvami, A.H. PRMA-Based Cognitive Machine-to-Machine Communications in Smart Grid Networks. *IEEE Trans. Veh. Technol.* **2015**, *64*, 3608–3623.
39. Bartoli, G.; Fantacci, R.; Letaief, K.; Marabissi, D.; Privitera, N.; Pucci, M.; Zhang, J. Beamforming for Small Cells Deployment in LTE-Advanced and Beyond. *IEEE Commun. Mag.* **2014**, *21*, 50–56.
40. Bhushan, N.; Li, J.; Malladi, D.; Gilmore, R.; Brenner, D.; Damnjanovic, A.; Sukhavasi, R.; Patel, C.; Geirhofer, S. Network densification: The dominant theme for wireless evolution into 5G. *IEEE Commun. Mag.* **2014**, *52*, 82–89.

41. Mazza, D.; Tarchi, D.; Corazza, G.E. A partial offloading technique for wireless mobile cloud computing in smart cities. In Proceedings of the European Conference on Networks and Communications (EuCNC), Bologna, Italy, 23–26 June 2014; pp. 1–5.

42. Zhou, L.; Hu, X.; Zhu, C.; Ngai, E.C.H.; Wang, S.; Wei, J.; Leung, V.C.M. Green small cell planning in smart cities under dynamic traffic demand. In Proceedings of the IEEE Conference on Computer Communications Workshops (INFOCOM WKSHPS), Hong Kong, China, 26 April–1 May 2015; pp. 618–623.

43. Huang, L.; Zhu, G.; Du, X. Cognitive femtocell networks: An opportunistic spectrum access for future indoor wireless coverage. *IEEE Wirel. Commun.* **2013**, *20*, 44–51.

44. Bu, S.; Yu, F.R. Green Cognitive Mobile Networks With Small Cells for Multimedia Communications in the Smart Grid Environment. *IEEE Trans. Veh. Technol.* **2014**, *63*, 2115–2126.

45. Augé-Blum, I.; Boussetta, K.; Rivano, H.; Stanica, R.; Valois, F. Capillary Networks: A Novel Networking Paradigm for Urban Environments. In Proceedings of the First Workshop on Urban Networking, Nice, France, 10–13 December 2012; pp. 25–30.

46. Shen, Z.; Papasakellariou, A.; Montojo, J.; Gerstenberger, D.; Xu, F. Overview of 3GPP LTE-advanced carrier aggregation for 4G wireless communications. *IEEE Commun. Mag.* **2012**, *50*, 122–130.

47. Al-Dulaimi, A.; Al-Rubaye, S.; Ni, Q.; Sousa, E. 5G Communications Race: Pursuit of More Capacity Triggers LTE in Unlicensed Band. *IEEE Veh. Technol. Mag.* **2015**, *10*, 43–51.

48. Yucek, T.; Arslan, H. A survey of spectrum sensing algorithms for cognitive radio applications. *IEEE Commun. Surv. Tutor.* **2009**, *11*, 116–130.

49. Leung, H.; Chandana, S.; Wei, S. Distributed sensing based on intelligent sensor networks. *IEEE Circuits Syst. Mag.* **2008**, *8*, 38–52.

50. Bansal, G.; Hossain, M.; Bhargava, V.; Le-Ngoc, T. Subcarrier and Power Allocation for OFDMA-Based Cognitive Radio Systems With Joint Overlay and Underlay Spectrum Access Mechanism. *IEEE Trans Veh. Technol.* **2013**, *62*, 1111–1122.

51. Lee, H.K.; Kim, D.M.; Hwang, Y.; Yu, S.M.; Kim, S.L. Feasibility of cognitive machine-to-machine communication using cellular bands. *IEEE Wirel. Commun.* **2013**, *20*, 97–103.

52. Ng, P. Optimization of Spectrum Sensing for Cognitive Sensor Network using Differential Evolution Approach in Smart Environment. In Proceedings of the IEEE 12th International Conference on Networking, Sensing and Control, Taipei, Taiwan, 9–11 April 2015; pp. 592–596.

53. Liu, F.; Wang, J.; Han, Y.; Han, P. Smart Grid Communication using Next Generation Heterogeneous Wireless Networks. In Proceedings of the IEEE Third International Conference on Smart Grid Communications, Tainan, Taiwan, 5–8 November 2012; pp. 229–234.

54. Tani, A.; Fantacci, R.; Marabissi, D. A low-complexity cyclostationary spectrum sensing for Interference Avoidance in femto-cells LTE-A based Networks. *IEEE Trans. Veh. Technol.* **2015**, *65*, 2747–2753.

55. Oh, D.; Lee, H.; Lee, Y. Cognitive radio based femtocell resource allocation. In Proceedings of the International Conference on Information and Communication Technology Convergence (ICTC), Jeju, Korea, 17–19 November 2010; pp. 274–279.

56. Sun, D.; Zhu, X.; Zeng, Z.; Wan, S. Downlink power control in cognitive femtocell networks. In Proceedings of the International Conference on Wireless Wireless Communications and Signal Processing (WCSP), Nanjing, China, 9–11 November 2011; pp. 1–5.

57. Yiu, S.; Chae, C.B.; Yang, K.; Calin, D. Uncoordinated Beamforming for Cognitive Networks. *IEEE Trans. Commun.* **2012**, *60*, 1390–1397.

58. Bartoli, G.; Fantacci, R.; Marabissi, D.; Pucci, M. LTE-A femto-cell interference mitigation with MuSiC DOA estimation and null steering in an actual indoor environment. In Proceedings of the IEEE International Conference on Communications (ICC), Budapest, Hungary, 9–13 June 2013; pp. 2707–2711.

59. Tarchi, D.; Fantacci, R.; Marabissi, D. Proposal of a cognitive based MAC protocol for M2M environments. In Proceedings of the IEEE 24th Annual International Symposium on Personal, Indoor, and Mobile Radio Communications (PIMRC), London, UK, 8–11 September 2013; pp. 1609–1613.

MDPI AG

St. Alban-Anlage 66

4052 Basel, Switzerland

Tel. +41 61 683 77 34

Fax +41 61 302 89 18

http://www.mdpi.com

*Future Internet* Editorial Office

E-mail: futureinternet@mdpi.com

http://www.mdpi.com/journal/futureinternet